American Airpower: Bombers 1935 – 1960

Vintage Photographs from WWII, Korea and the Cold War

Author: John Cilio

Editors: Herb Hill & Lisa Maxwell

Vintage Flyer Media LLC

Printed in the U.S.A.

Front cover photograph: A 20th Bomber Command B-29 bomber was photographed while ground crews loaded 500 pound bombs. The base, somewhere in India, was preparing its bombers for a raid on an air depot at Okyama. With a conventional bomb load of 20,000 pounds or one nuclear weapon, the bomber packed a punch. The last B-29 bomber was completed by Boeing's Renton factory on May 28, 1946. (U.S. Army Air Forces)

Cilio, John, 2011.

American Airpower: Bombers 1935-1960 Vintage Photographs of WWII, Korea and the Cold War / John Cilio

ISBN: 978-0-9827728-7-4

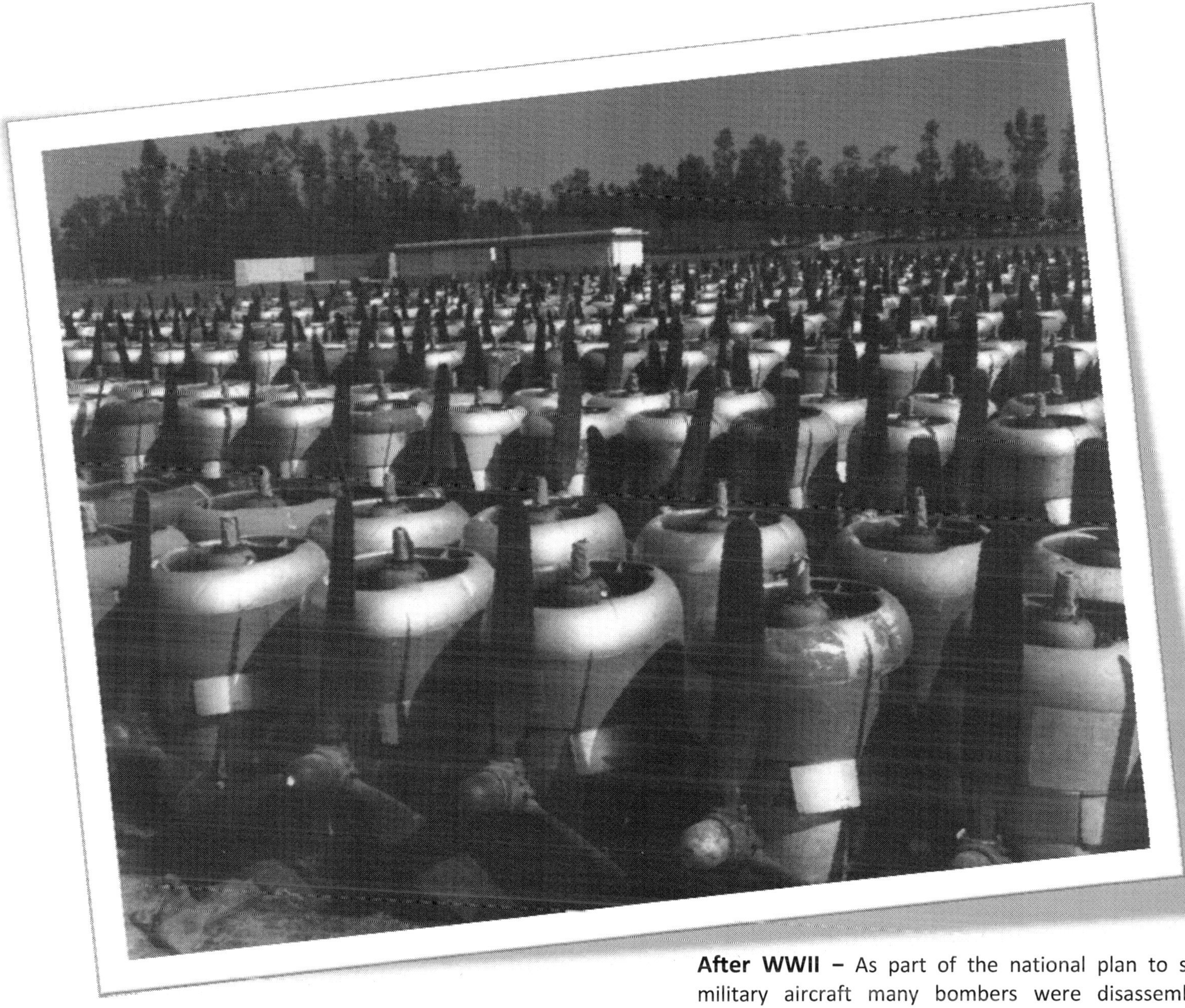

After WWII – As part of the national plan to scrap military aircraft many bombers were disassembled. These engines were photographed standing at attention while they awaited their fate. (Private Collection)

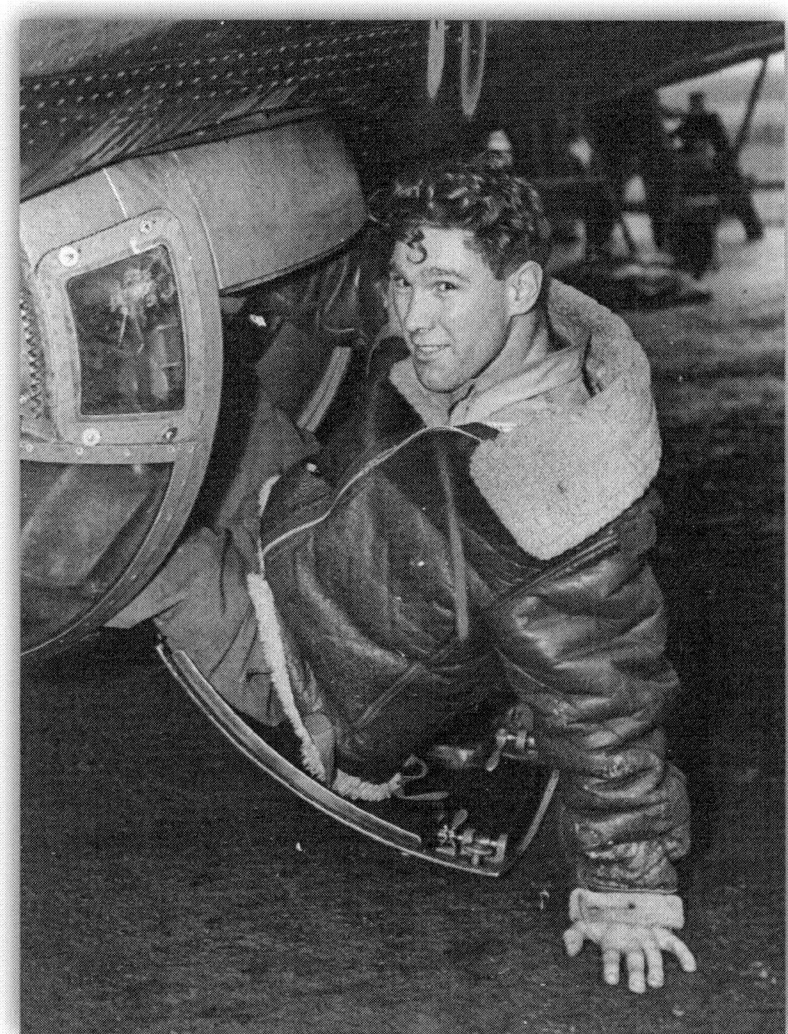

Ball Turret Gunner – Staff Sergeant Glenn Allen, of Albion, NY, (left) was awarded the Air Medal and three oak leaf clusters at the time of this photograph. He was exiting a ball turret from a B-17. He was 22, engaged to Betty Scheuerer, and looking forward to getting home. (British Combine)

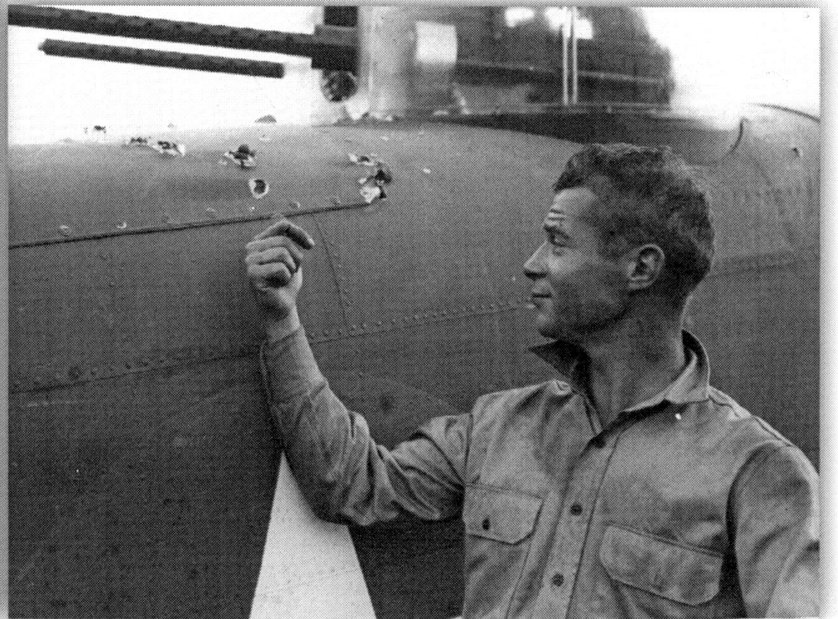

Top Turret Gunner – Sergeant Jack Gould, from San Francisco, CA, was photographed while he explained the bullet holes in his B-25 which were put there by an attacking Japanese Zero during a raid over Burma. He said, "He was trying to get me when I got him." (U.S. Army Air Forces)

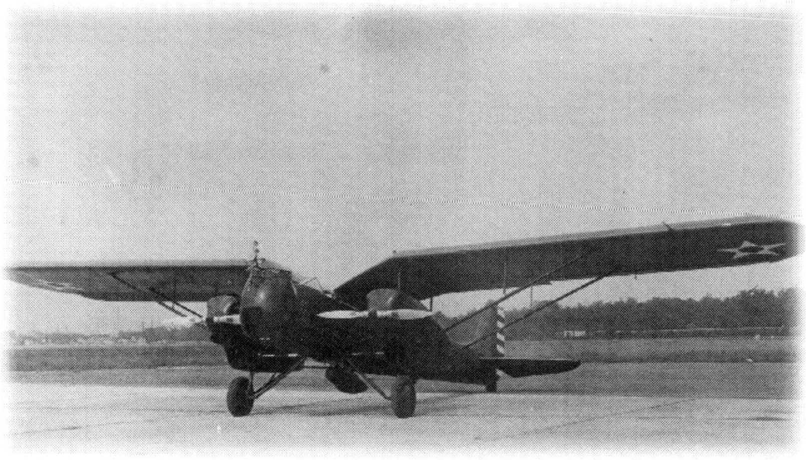

Douglas Y1B-7 – The Douglas Y1B-7 was the first monoplane bomber to receive the bomber (B) designation. It was considered a light bomber, carrying only 1,200 pounds of bombs. It carried a crew of four: pilot, copilot, nose gunner and rear gunner. (U.S. Air Force Museum)

An Introduction to *American Airpower: Bombers 1935-1960*

American Airpower: Bombers 1935-1960, brings together a twenty-five year collection of aircraft whose military role was to drop explosives on a target from the air. The photographs represent American bombers that flew in WWII, the Korean War and the early Cold War. It was a time when computer power was negligible yet engineers dramatically improved bomber capabilities, payloads and speed. In 1935, America's largest bombers could approach 200 mph. Twenty-five years later, the Cold War B-58 bomber could fly almost twice the speed of sound, at 1,320 mph.

Over this relatively short time span, several different bombers were required to deliver the full punch of American airpower. Until the late 1940s, bombers were pretty much categorized based upon their bomb capacity. They were heavy, medium or light bombers. Other bomber classifications were based upon the type of attack that the aircraft was capable of completing. These aircraft were known as dive bombers and fighter-bombers. As the Cold War evolved, bomber designations expanded to include new terminology that recognized evolving military strategies. Bombers in addition to being designated as heavy or medium, as an example, were also classified as tactical, strategic, conventional or nuclear.

The magnitude of the destruction caused by bombing in WWII was unimaginable. As the Cold War escalated, the danger of entire cities being annihilated by bombers became a real possibility. Fortunately, bomber strategy changed. Instead of delivering destruction, bombers became the primary deterrent to conflict because of the damage they could inflict. One Cold War bomber, the B-36, was even called the Peacemaker to emphasize the point. Eventually, intercontinental missiles would share the deterrent responsibility.

As you turn the pages of this book, you will see photographs of all of the bombers that were used in action, gain a sense of the volume of aircraft required in war and gain a perspective of how the crews and their ground eco-systems contributed to the success of each bomber mission. American airpower remains strong, partly because of the valor that the crews and aircraft depicted within these pages delivered in service to their country.

An initiation to bomber aircraft evolution –

Dropping explosives from the air has been used as a military tactic since the evolution of balloons and aircraft capable of carrying the extra weight of explosives. Added weight meant less maneuverability until aircraft improved and more weight could be carried without significantly impacting aircraft performance. Some of the largest bombers of WWI were German dirigibles. They bombed London, flying higher than most aircraft or anti-aircraft weapons could reach. Their bombs terrorized the cities they fell upon.

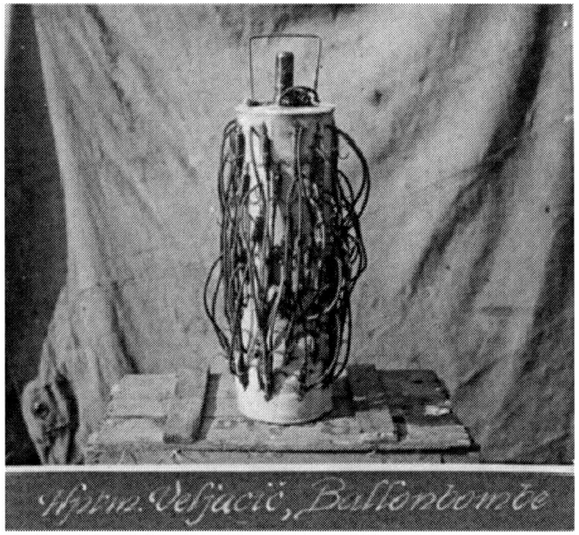

A WWI German aerial bomb that could be dropped by dirigible, photographed about 1917. (Author's private collection)

The first aircraft in the U.S. to officially be designated a bomber was the XB-7. Designed by Douglas Aircraft in 1930/1931, the aircraft's speed and range enhancements foreshadowed the end of the biplane bomber. Carrying bomb racks mounted outside the fuselage, it was capable of carrying up to 1,200 lbs of bombs 400 miles. The XB-7 aircraft never went into mass production. During the time it was in development, larger, more powerful bombers were on the drawing board.

Martin led the development effort and created the B-10. It was to become the first mass produced bomber whose performance was superior to that of the Army's pursuit aircraft (fighters) of the time.

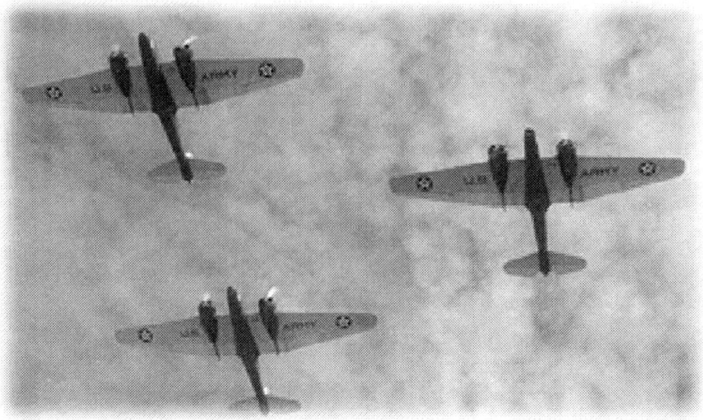

Martin B-10 bombers (U.S. Army Air Force)

The B-10, with a 70 foot wingspan, entered service in 1934.[1] It featured all-metal construction, enclosed cockpits, rotating gun turrets, retractable landing gear, an internal bomb bay, and full engine cowlings. It set the direction of bombers for the next several decades. Bombers continued to evolve and soon Douglas and Boeing Aircraft introduced designs that leveraged the new standards in design: the Boeing B-15, the Douglas B-18 and revolutionary Boeing B-17.

The B-15 was a valuable test bed for the future but its design was ahead of engine technology and it was too slow to merit production. The B-18 was also slow and carried a very light bomb load inadequate for the impending war effort. The B-17 was America's largest bomber at the start of WWII. Other designs followed. The B-24 was about the same size as the B-17 but had larger bomb capacity and could deliver them to more distant targets, faster. More powerful bombers continued to evolve: the B-29, the B-36 and the B-45 with jet engines. Technology and the strategies to leverage their power continued to advance. Today's bombers, manufactured by many of the same companies as in WWII, share the same mission as their predecessors, to deliver explosives by air, but now they deliver their payloads with smart bombs and stealth proficiency.

[1] Eden, Paul and Soph Moeng, eds. *The Complete Encyclopedia of World Aircraft.* London: Amber Books Ltd., 2002.

Big Bombers – The Boeing B-15

APPROVED BY WAR DEPARTMENT

FROM: BOEING NEWS BUREAU
Boeing Aircraft Company
Seattle 14, Washington

THROUGH: N. W. AYER & SON, INC.
30 Rockefeller Plaza
New York 20, N. Y.

For Immediate Release

6/14/45

BOEING'S "GRANDPAPPY" POSES FOR LAST TIME

SEATTLE, Wash. -- Long past the "three score and ten" of his airplane brotherhood, the history-making Boeing B-15 has at long last been taken from service and is headed for retirement, Boeing Aircraft Company officials learned this week.

News of the last mission of the solitary B-15, first of the super-bombers and "foster-father" of the Boeing B-29 Superfortresses which are roaming the skies over Japan, came to Boeing a little more than eight years after the big airplane rolled out from the local plant.

Since that day, April 17, 1937, the B-15, first of the really big U. S. bombers, has made a name for himself by one spectacular performance after another.

"Grandpappy," as the big bomber became known in the Air Forces, had a wing spread of 150 feet, nearly nine feet greater than that of the B-29. Built two years after the first Boeing B-17 Flying Fortress, the B-15 was severely handicapped because at that time there were no engines powerful enough to let him perform as he otherwise could have.

But "Grandpappy" made history in the skies. On February 4, 1939, he took off from Langley Field, Va., carrying medicines and supplies, plus an 11-man crew, for an earthquake-stricken area in Chile. A little more than 50 hours later, the airplane settled down at Santiago, Chile, having made intermediate landings at Colon, C. Z. and Lima, Peru. Returning from the mercy mission, "Grandpappy" flew 3,000 miles non-stop.

Later that year he broke all existing records by carrying a payload of 31,205 pounds, plus his own hefty bulk of 43,000 pounds, to a height of 8,200 feet. This was 1,638 feet and 2,545 pounds better than the previous world record established by a Russian airplane.

(Original Press Release)

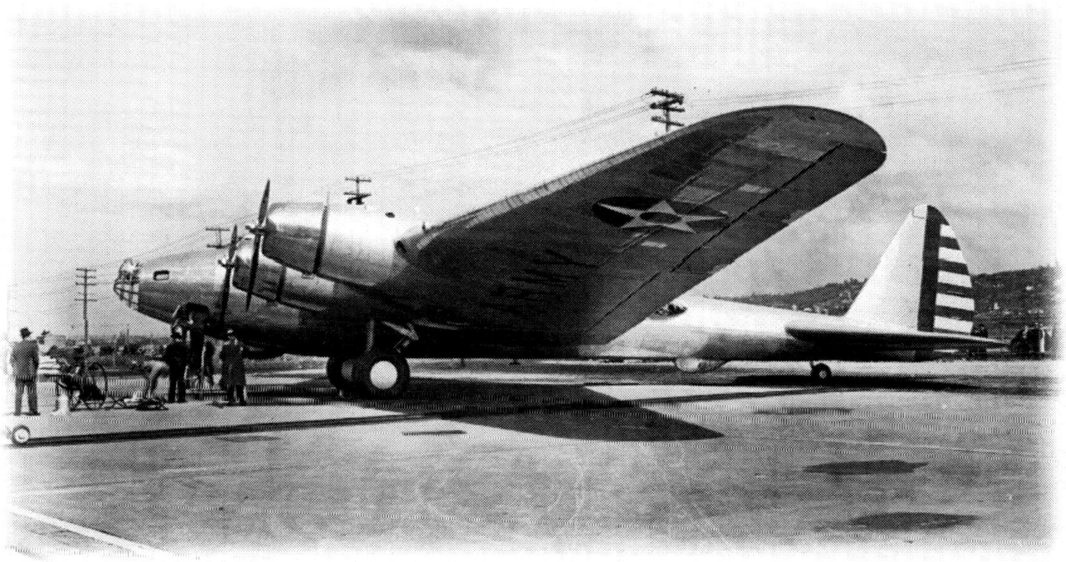

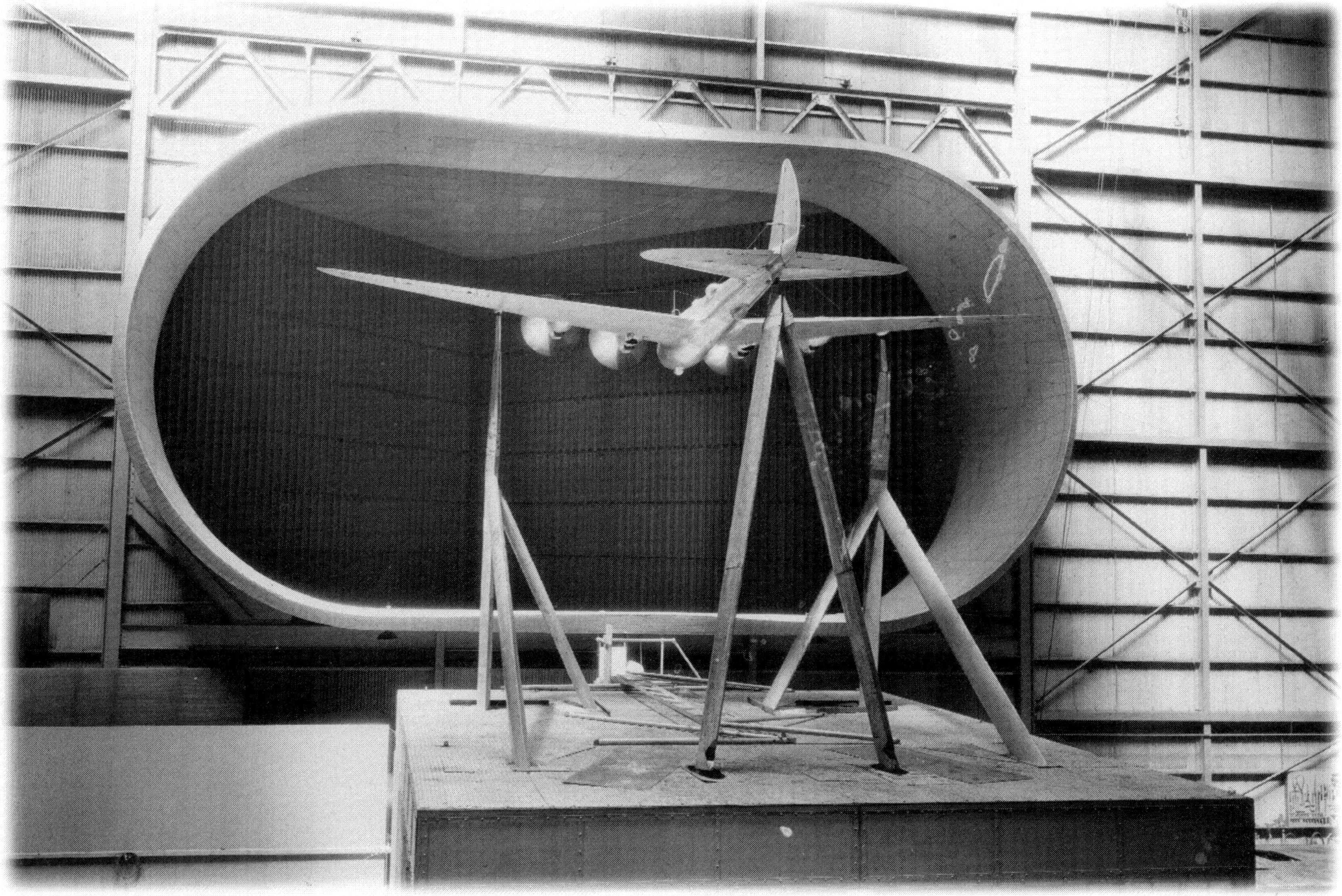

Designing a Bomber – This mid-1930s photograph shows a YB-17 full scale model with a 103 foot wingspan being tested in the wind tunnel at the National Advisory Committee for Aeronautics. WWII aircraft designers didn't have the luxury of computer simulation; instead, they tested their designs through model mock-ups. Smaller scale models were initially used and as the aircraft matured, full scale models were built and studied under the strain of wind tunnel simulation. The YB-17 was a descendant of an earlier Boeing 299 experimental aircraft which crashed. Eventually 13 YB-17s were built and all but one were assigned to the 2nd Bomb Group based at Langley Field, VA. The 13th YB-17 was the only aircraft actually used for extensive flight testing and helped develop heavy bomber techniques before WWII.[1]
(National Advisory Committee for Aeronautics) [1]National Museum of the U.S. Air Force

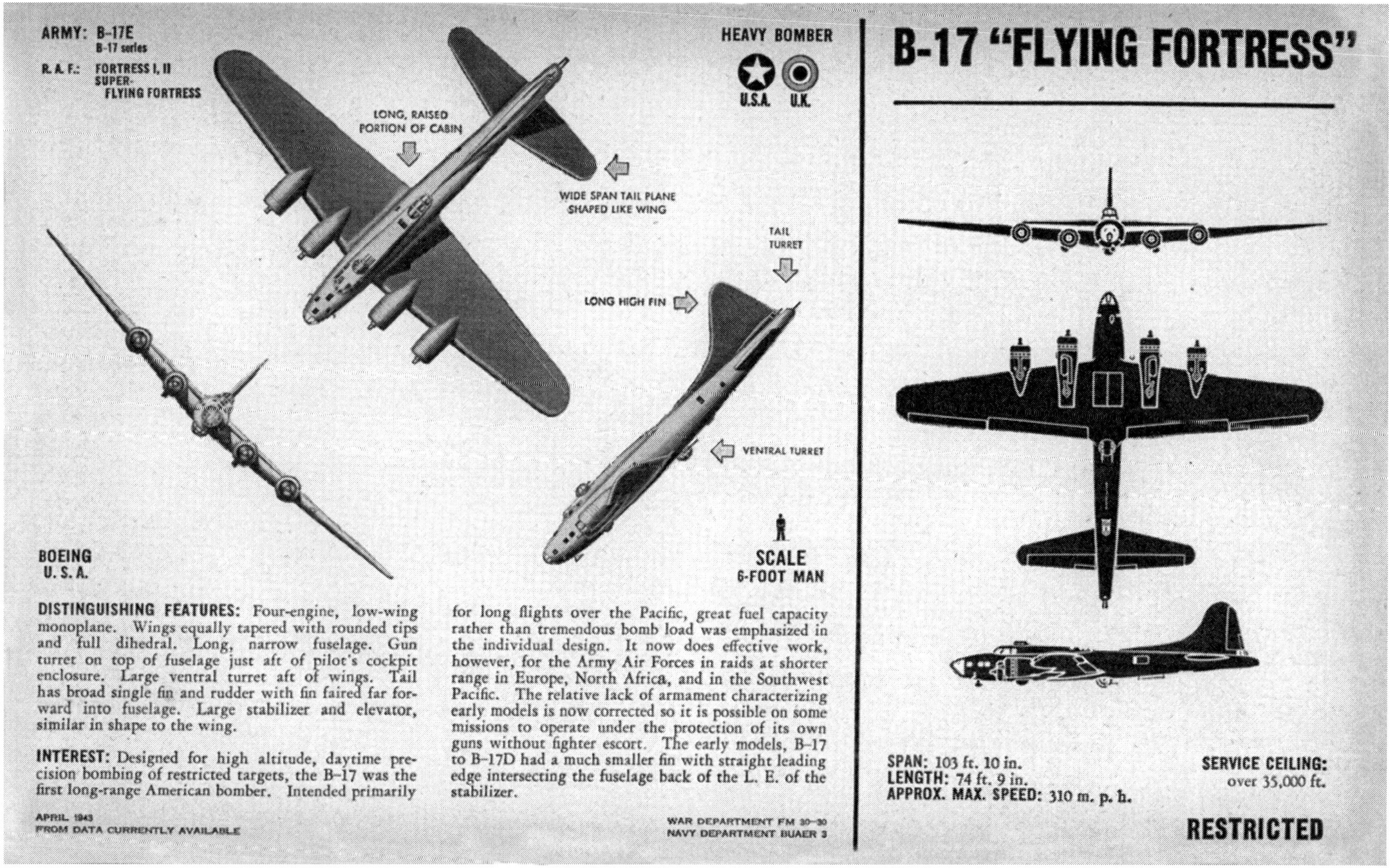

ARMY: B-17E
B-17 series

R. A. F.: FORTRESS I, II
SUPER-
FLYING FORTRESS

HEAVY BOMBER

U.S.A. U.K.

B-17 "FLYING FORTRESS"

BOEING
U. S. A.

SCALE
6-FOOT MAN

DISTINGUISHING FEATURES: Four-engine, low-wing monoplane. Wings equally tapered with rounded tips and full dihedral. Long, narrow fuselage. Gun turret on top of fuselage just aft of pilot's cockpit enclosure. Large ventral turret aft of wings. Tail has broad single fin and rudder with fin faired far forward into fuselage. Large stabilizer and elevator, similar in shape to the wing.

INTEREST: Designed for high altitude, daytime precision bombing of restricted targets, the B-17 was the first long-range American bomber. Intended primarily for long flights over the Pacific, great fuel capacity rather than tremendous bomb load was emphasized in the individual design. It now does effective work, however, for the Army Air Forces in raids at shorter range in Europe, North Africa, and in the Southwest Pacific. The relative lack of armament characterizing early models is now corrected so it is possible on some missions to operate under the protection of its own guns without fighter escort. The early models, B-17 to B-17D had a much smaller fin with straight leading edge intersecting the fuselage back of the L. E. of the stabilizer.

SPAN: 103 ft. 10 in.
LENGTH: 74 ft. 9 in.
APPROX. MAX. SPEED: 310 m. p. h.

SERVICE CEILING: over 35,000 ft.

APRIL 1943
FROM DATA CURRENTLY AVAILABLE

WAR DEPARTMENT FM 30-30
NAVY DEPARTMENT BUAER 3

RESTRICTED

B-17 Flying Fortress – The Boeing B-17 Flying Fortress was a heavy bomber used by the U.S. Army Air Forces in all theatres of WWII. The image above is from a restricted use military spotter manual. Often these images were mixed with other aircraft, including enemy aircraft, and shown as flash cards to military personnel during training. The drills helped personnel learn to quickly decide to shoot or not shoot at an aircraft. Early versions of the B-17 proved to be more vulnerable to fighter attack than anticipated. Improvements were made and before the United States entered WWII in 1941, the B-17E version was equipped with more powerful engines and gun turrets in the upper fuselage, belly, and tail. Additional enhancements were made during the war. (War Department Recognition Guide 1943)

B-17 Flying Fortress

Esmeralda was the name of this particular B-17 but the name "Flying Fortress" was universal to the aircraft. Its name was not part of any plan but was coined when one of the reporters present during the aircraft's unveiling at the Boeing plant, remarked, "It looks like a 15-ton flying fortress".[1]

Although the B-17 was just 19 feet tall, the scale of a guard standing next to the 36,000 pound aircraft is thought provoking. In time it became a small bomber when compared to the bombers in use at the end of WWII.

The name Flying Fortress stayed with the bomber during its entire WWII service. The aircraft accounted for over 290,000 sorties against ground installations and dropped over 640,000 tons of bombs.[2] Just over 12,700 were built and at times during the war, particularly when no fighter escort was available, it suffered significant casualty rates.

(U.S. Air Forces)

[1&2] www.militaryfactory.com

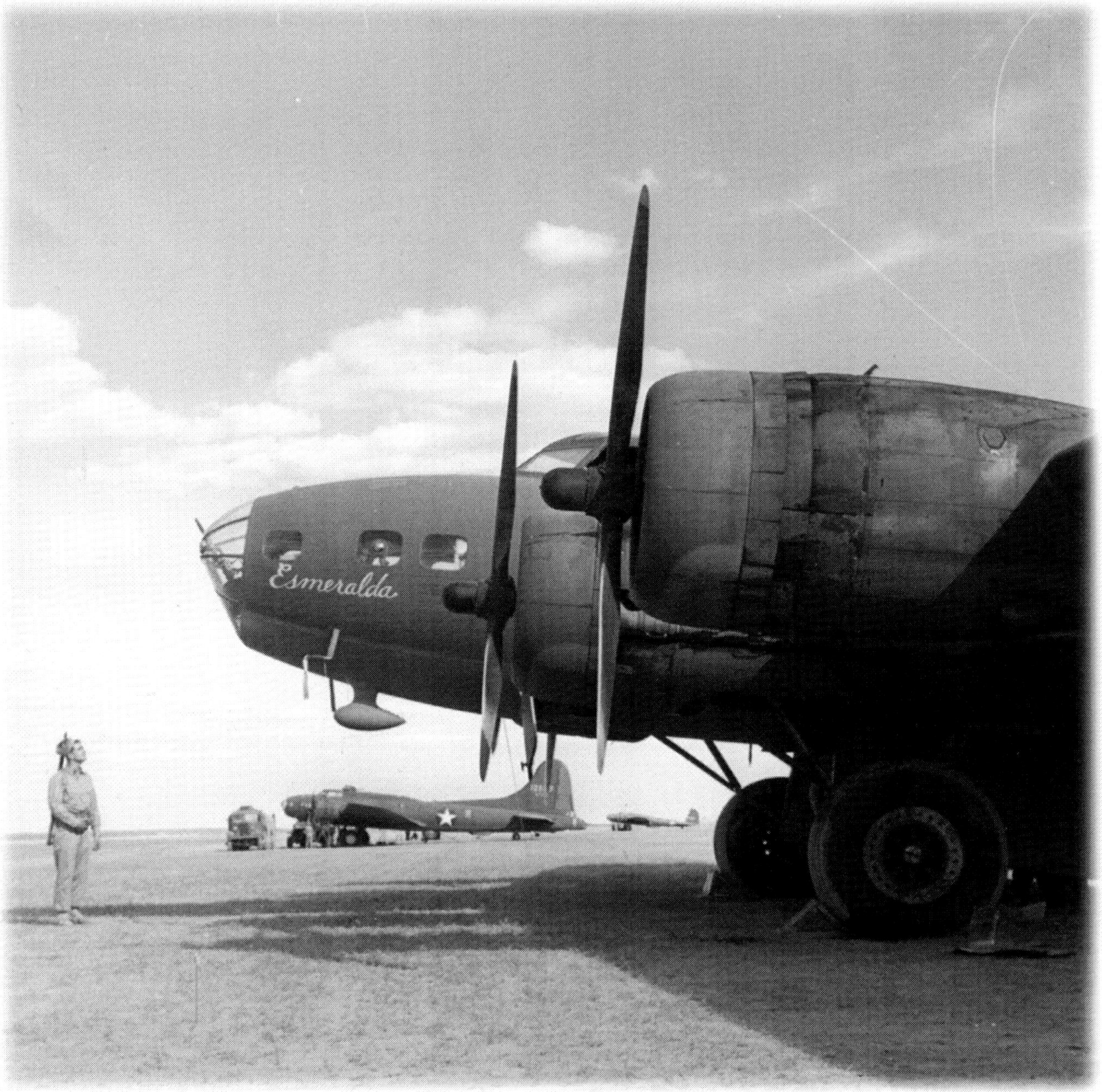

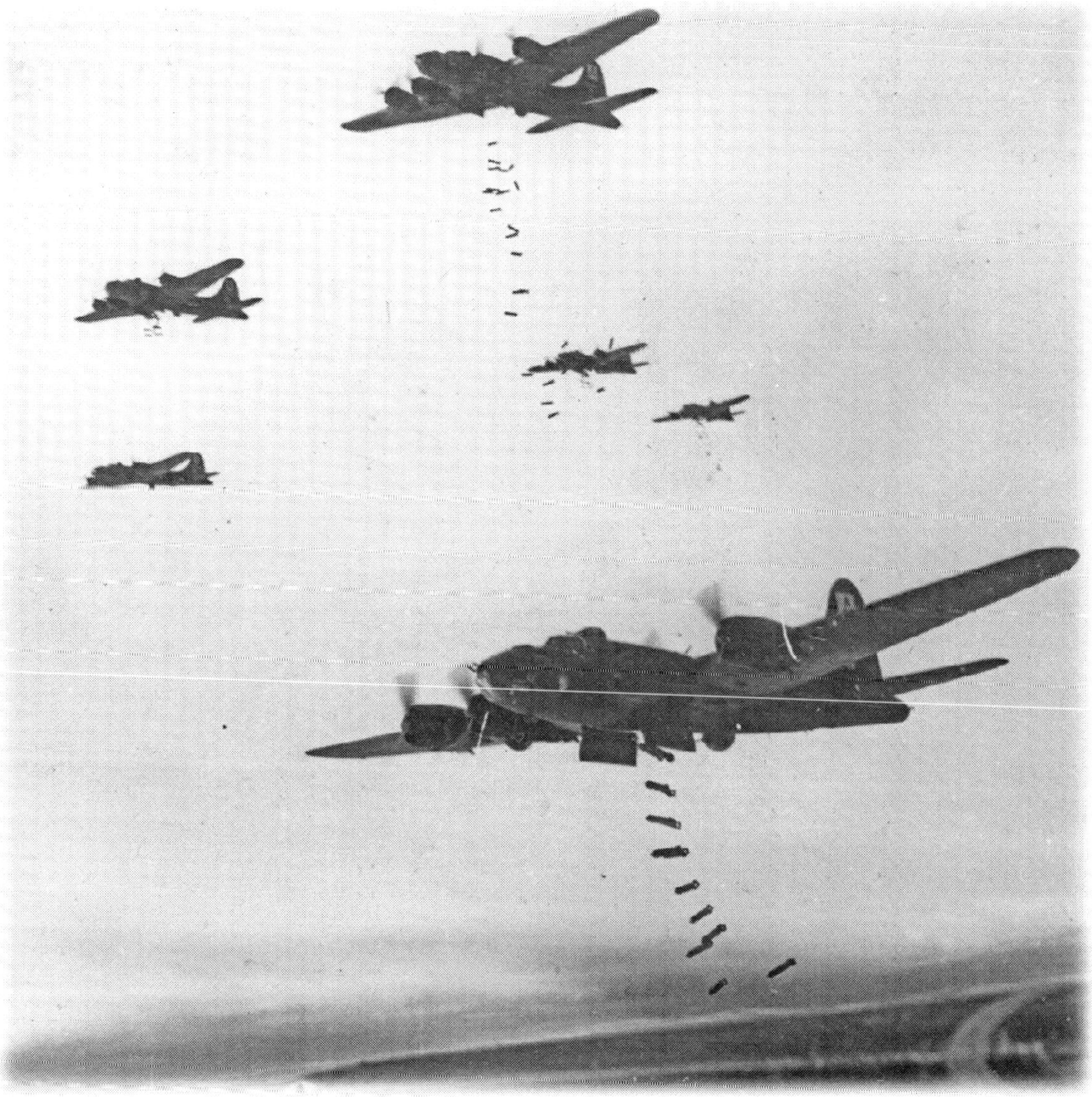

B-17 Flying Fortress

The Flying Fortress was an important bomber in Europe but also fought in the Pacific Theatre from December 7, 1941, until its last Pacific combat mission on October 20, 1943.[1] The bomber was designed to attack strategic targets deep within enemy territory through precision daylight bombing. The aircraft would penetrate the enemy territory by flying above the effective range of anti-aircraft artillery with its turbo-supercharged radial engines. The engines provided the necessary high-altitude performance, and heavy defensive armament provided moderate protection against attacking fighters.

When enough bombers were available they flew in tightly stacked defensive formations. The typical formation was called a squadron "box" of nine or 12 aircraft. When there were three squadron boxes, staggered vertically and horizontally, it was called a group, and three groups in a trail formed a combat wing. The formations offered mutual protection from their gunners. Early B-17s were not equipped with enough frontal firepower and the weakness was exploited by the enemy. The B-17 loss rate exceeded 25% on numerous early missions.[2]

(U.S. Army Air Corps) [1]*Fortress Against the Sun: The B-17 Flying Fortress in the Pacific*. By Gene Eric Salecker ISBN: 1-58097-049-4
[2]Caidin, Martin. *Black Thursday*. New York: E.P. Dutton & Company, 1960. ISBN 0-553-26729-9.

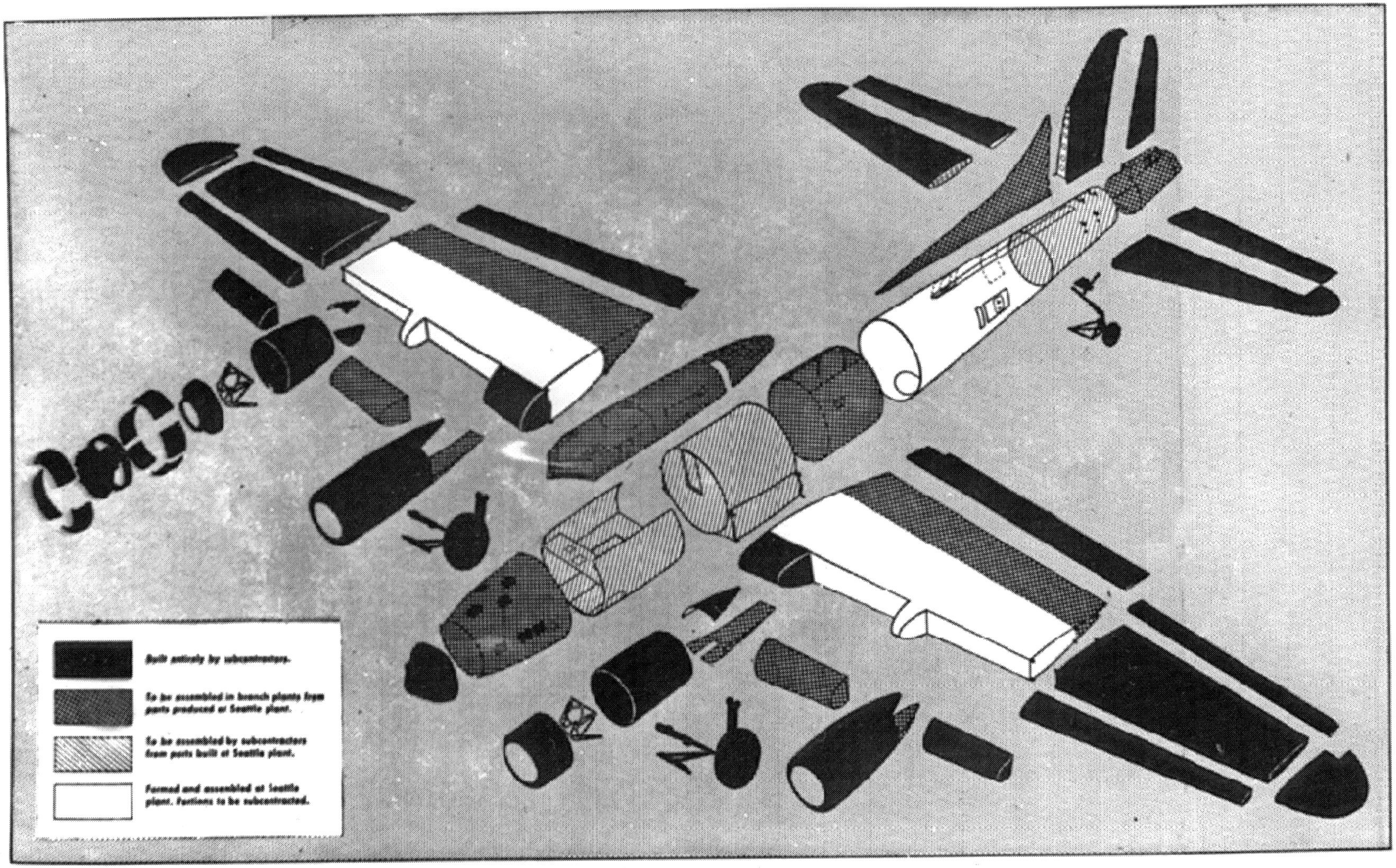

B-17 Flying Fortress – Up until WWII, most aircraft manufacturers had not produced large volumes of aircraft. In 1940, Boeing had begun to anticipate the requirement of mass bomber production. Their retooling plan would no longer build aircraft one at a time, instead multiple aircraft would be assembled simultaneously on an assembly line. Their planning paid dividends because after Pearl Harbor, the company was given the direction to produce as many B-17 bombers as possible. Building the huge bomber (74 feet long with an almost 104 foot wingspan) and all of its subassemblies was a production space challenge. The image above shows how Boeing maximized B-17 production by sub-contracting work outside the main plant and eventually multiple plants. The areas in black were built using sub-contractors to build parts outside of Boeing, parts shown as dark shaded areas were subassemblies from Boeing plants made outside of Seattle and the cross-hatched parts represent sections of the plane assembled by sub-contractors from parts made in Seattle. All of the pieces were then centralized and assembled with massive jigs. (Boeing Airplane Company)

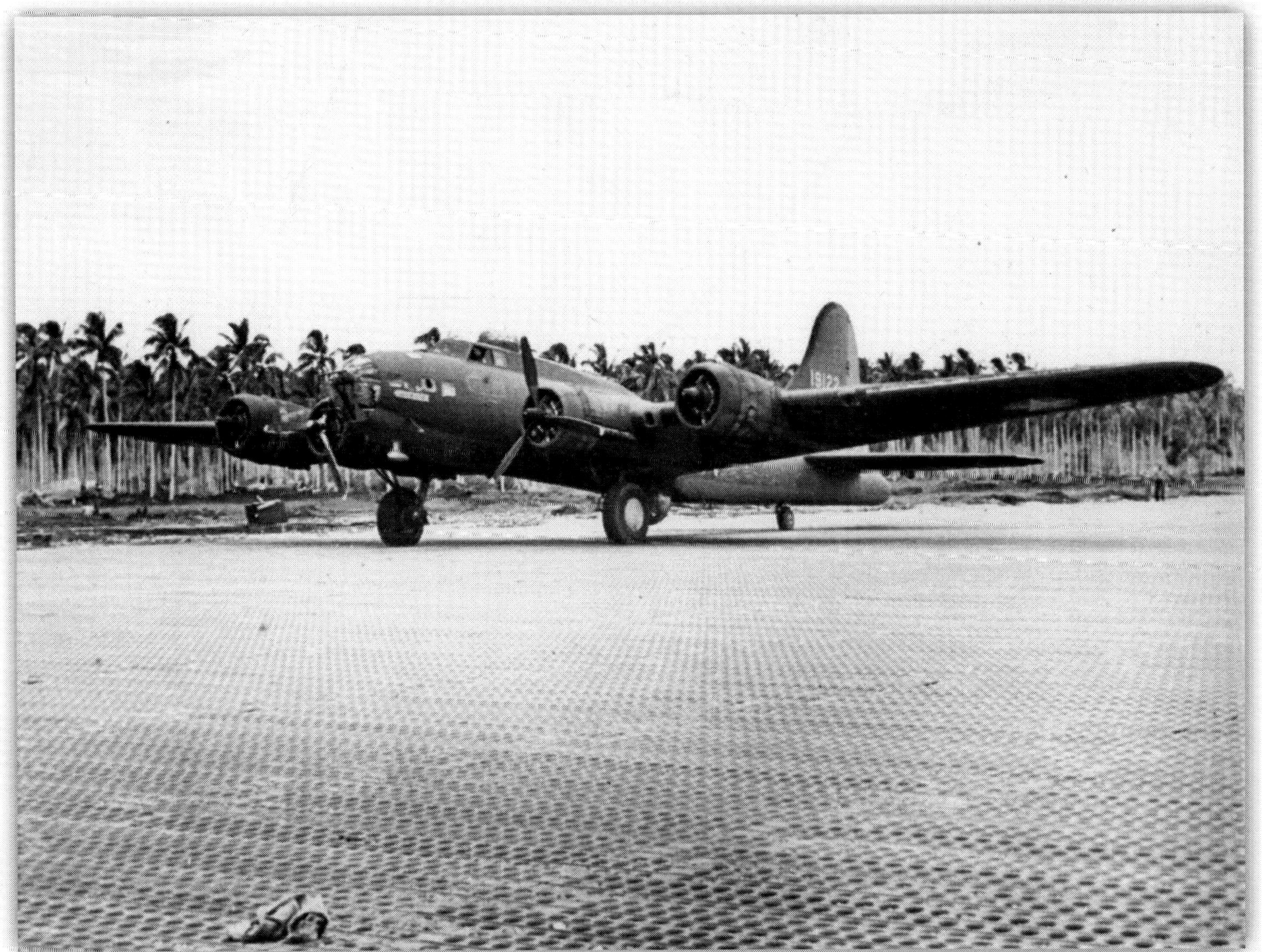

B-17 Flying Fortress – This B-17 was photographed parked at the Guadalcanal Airport shortly after U.S. forces had taken the airport from the Japanese. It was parked on a U.S. Army Corps' innovation – steel mats known as the Marston Mat. The mats were laid in small sections, 16" by 10', which could be quickly rolled out to form a runway surface to facilitate takeoffs and landings on sandy soils. The B-17 was not used extensively in the Pacific because of the limited range for Pacific Theatre tactics and the lack of sufficient quantities to create mass formations for raids. (Wide World Photo)

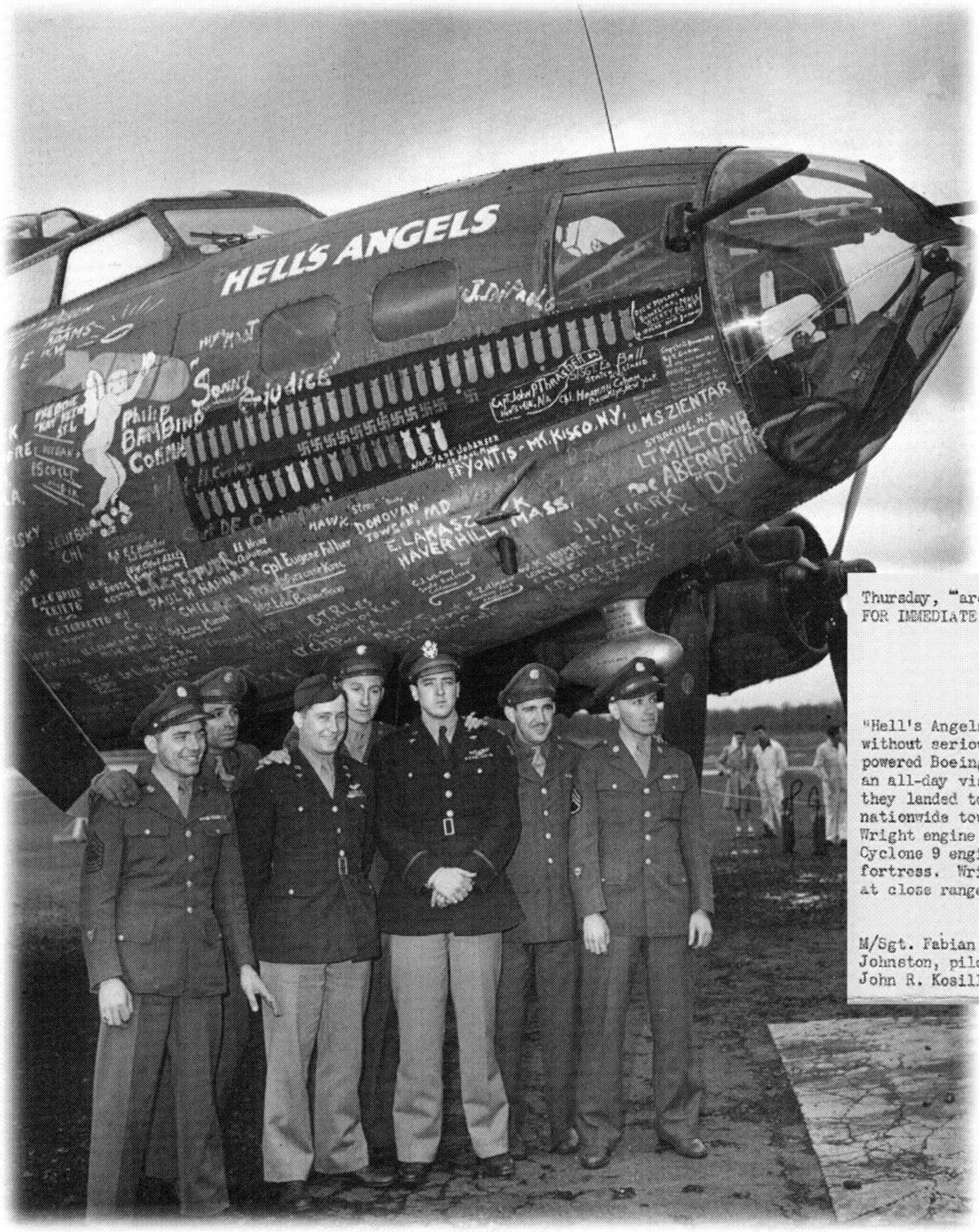

Celebrating Twenty-five Successful Missions

On May 13, 1943, *Hell's Angels*, the B-17F from the 303rd Bomb Group, became the first 8th Air Force B-17 to complete 25 combat missions.

When the aircraft completed 48 combat missions on December 13, 1943, it retired from combat and returned to the U.S. The crew had adopted the name of the aircraft from the 1927 WWI fictional Fighter Squadron movie, "*Hell's Angels*," by Howard Hughes.
(U.S. Army Air Corps)

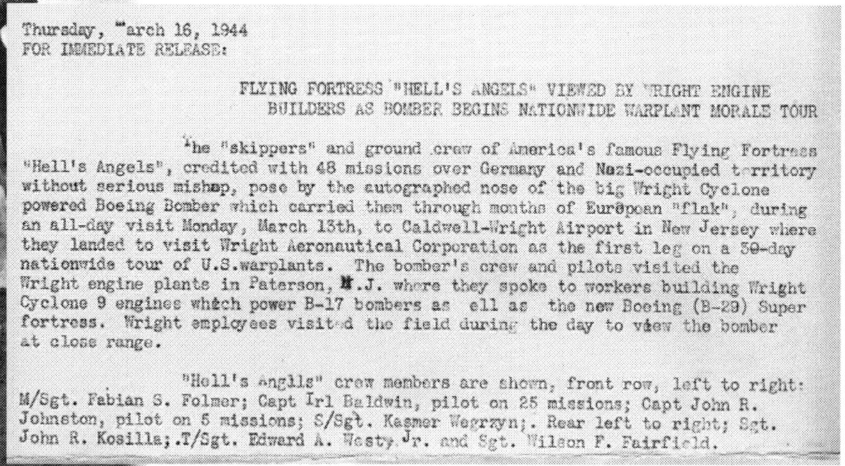

Thursday, March 16, 1944
FOR IMMEDIATE RELEASE:

FLYING FORTRESS "HELL'S ANGELS" VIEWED BY WRIGHT ENGINE BUILDERS AS BOMBER BEGINS NATIONWIDE WARPLANT MORALE TOUR

The "skippers" and ground crew of America's famous Flying Fortress "Hell's Angels", credited with 48 missions over Germany and Nazi-occupied territory without serious mishap, pose by the autographed nose of the big Wright Cyclone powered Boeing Bomber which carried them through months of European "flak", during an all-day visit Monday, March 13th, to Caldwell-Wright Airport in New Jersey where they landed to visit Wright Aeronautical Corporation as the first leg on a 30-day nationwide tour of U.S.warplants. The bomber's crew and pilots visited the Wright engine plants in Paterson, N.J. where they spoke to workers building Wright Cyclone 9 engines which power B-17 bombers as well as the new Boeing (B-29) Super fortress. Wright employees visited the field during the day to view the bomber at close range.

"Hell's Angils" crew members are shown, front row, left to right: M/Sgt. Fabian S. Folmer; Capt Irl Baldwin, pilot on 25 missions; Capt John R. Johnston, pilot on 5 missions; S/Sgt. Kasmer Wegrzyn;. Rear left to right; Sgt. John R. Kosilla;.T/Sgt. Edward A. Westy Jr. and Sgt. Wilson F. Fairfield.

(Original 1944 Press Release)

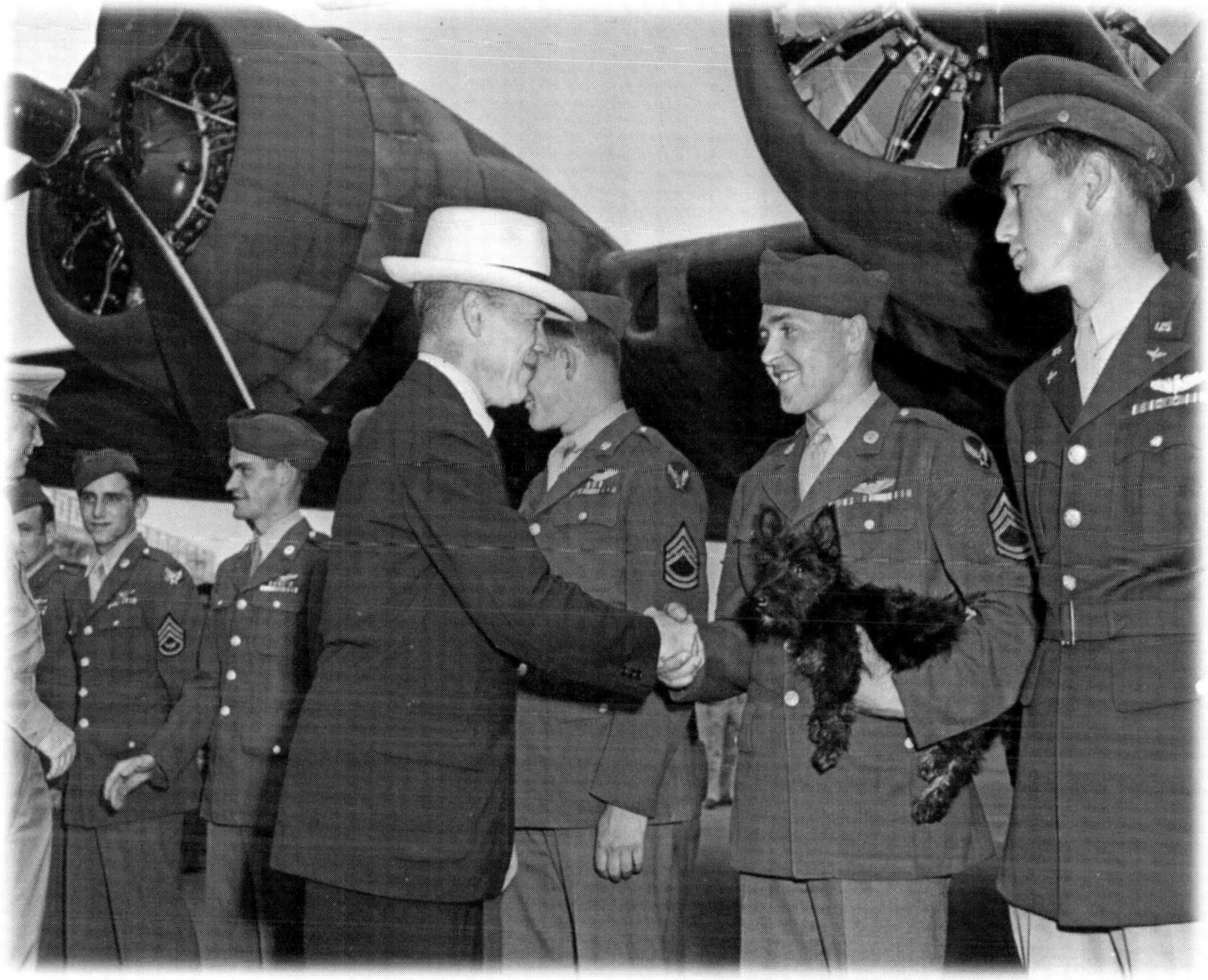

B-17 Flying Fortress *Memphis Belle* Crew and Mascot – The *Memphis Belle* was the second B-17 in the 8th Air Force to complete 25 missions doing so on May 19, 1943. It was the first to complete 25 missions with its crew intact. It was also the first to return to the U.S. and received an official welcome home at Washington D.C.'s National Airport. Shown on the far left, General H.H. Arnold, Chief of the Army Air Forces, shook hands with the crew. In the center with white hat, the Assistant Secretary of War, Robert Patterson, shook hands with Sergeant H.P. Lock who holds the *Memphis Belle* mascot, Stuka. (U.S. Air Forces)

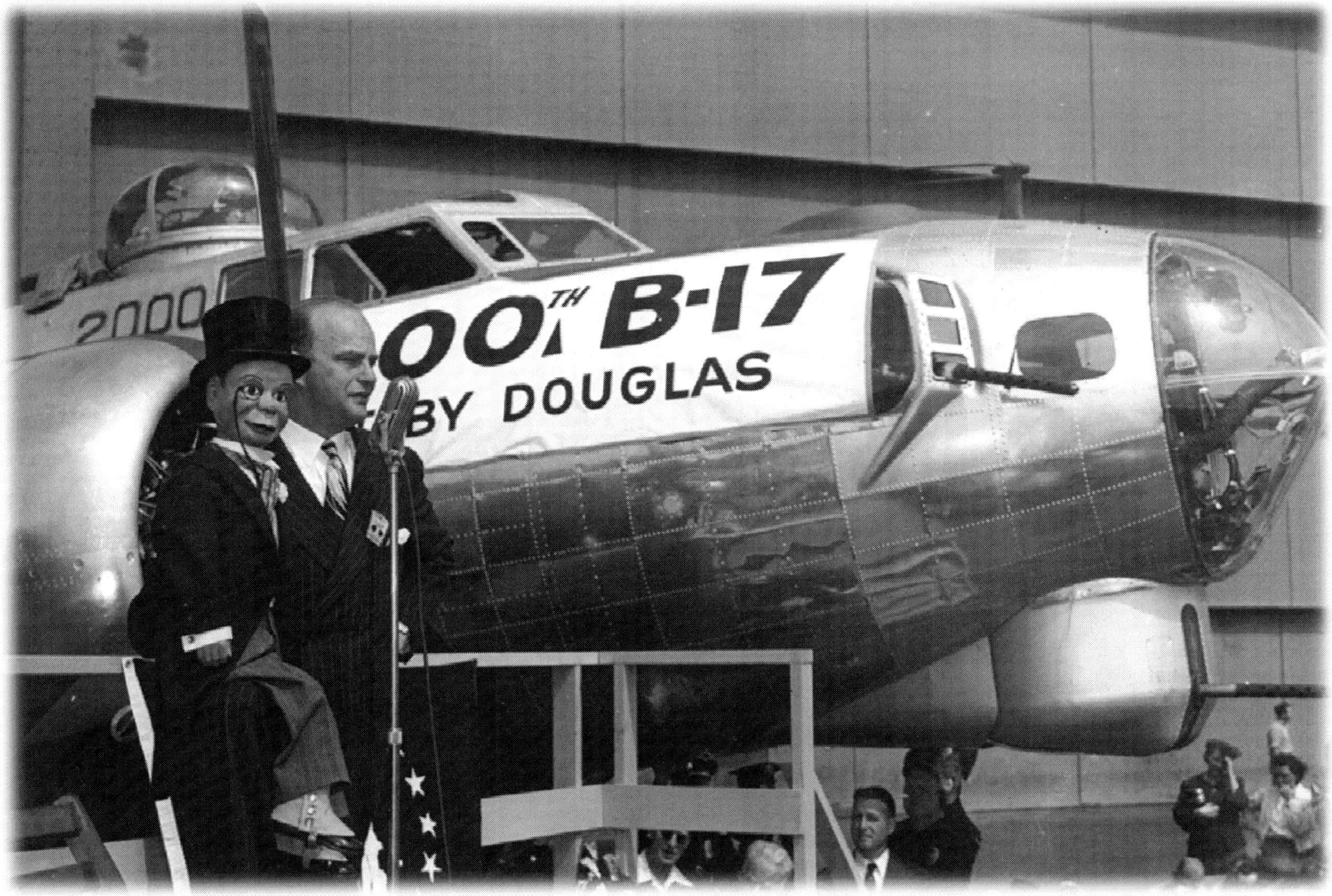

2,000th B-17 Flying Fortress – This 1942 photograph shows part of the celebration that surrounded the completion of the 2,000th B-17 bomber. The celebration featured Charlie McCarthy and Edgar Bergen, famous entertainers at the time. More than 10,000 more B-17s would be produced during WWII at three manufacturing facilities. Production was one way to defeat the enemy and a key focus of American aviation during the war. The 10,000th bomber celebration was held just two years later on October 9, 1944. Talking with WWII German pilots, they commented that their leaders said that America's outlandish claims of aircraft production were purely for publicity and propaganda and not possible. The announcements may have been for publicity and propaganda but the aircraft were in fact flying into battle. (Boeing and Consolidated Aircraft Corporation)

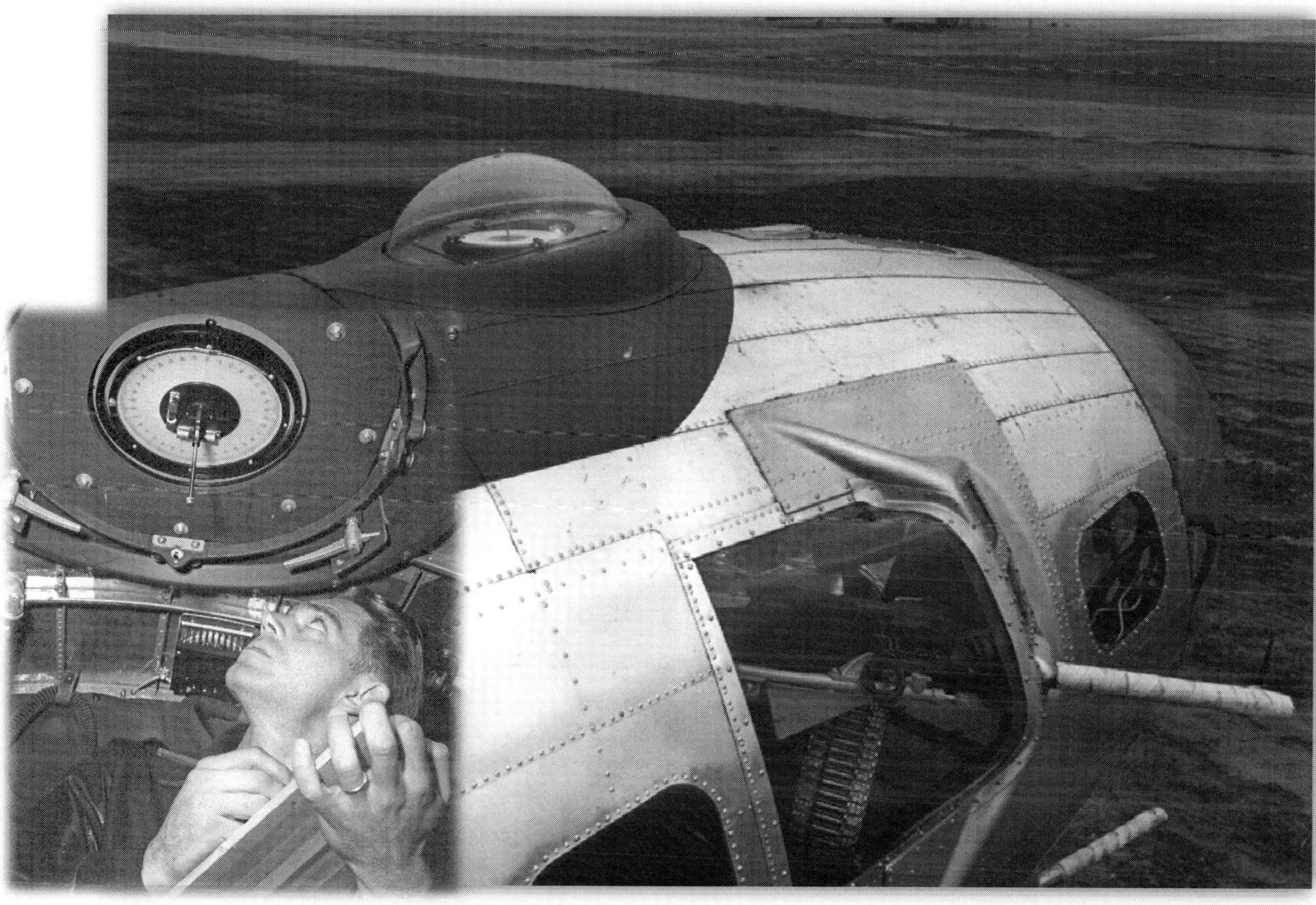

B-17 Flying Fortress – Boeing developed many improvements to the B-17 during WWII. Shown above is the introduction of a new compass, called a shadow compass. The new design was important because a magnetic compass could be influenced from enemy subversive efforts. No outside influences affected the shadow compass. The compass was installed in the astrodome on top of the fuselage in a painted area. Six areas on the B-17 were painted to keep glare from the sun out of the eyes of crew members. The inset photograph shows how the crew used the compass to take a bearing inside the aircraft. (Both photographs by Boeing Airplane Company)

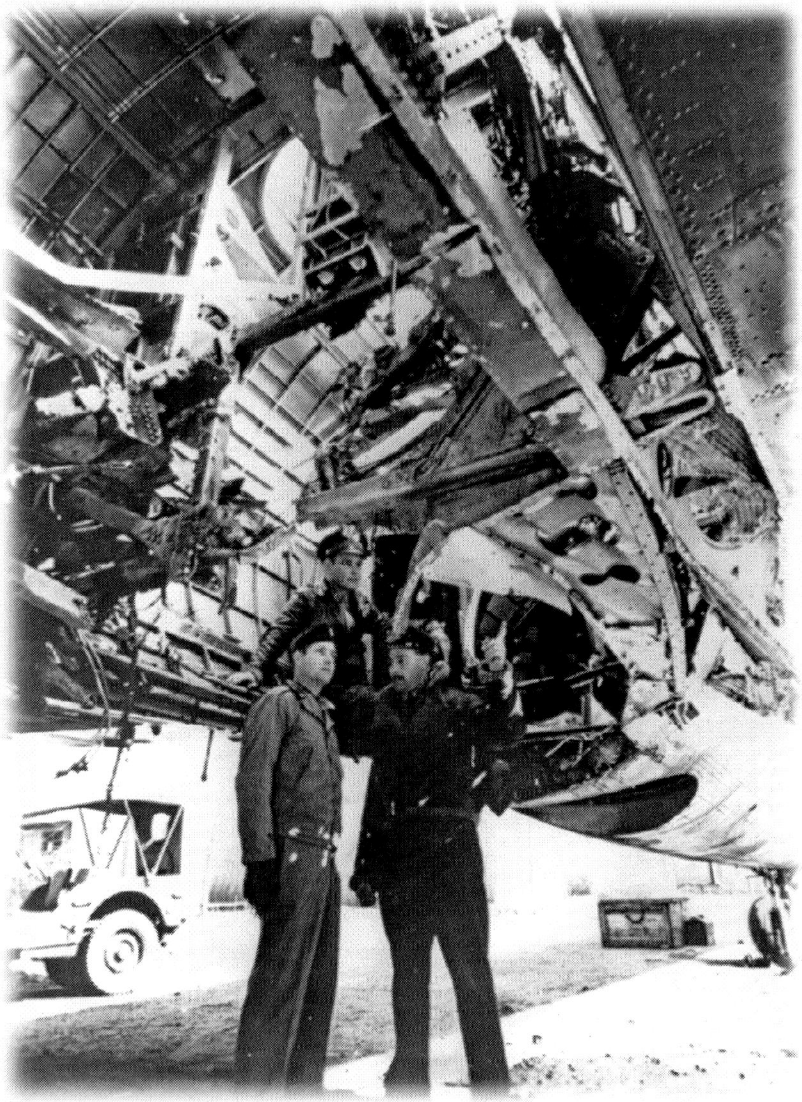

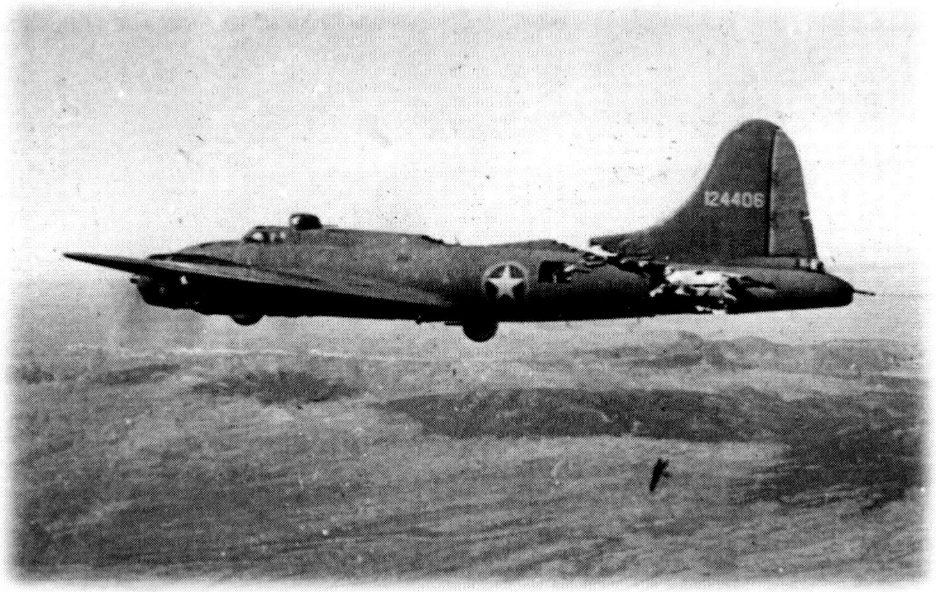

FORT A "FLYING MIRACLE"

This photograph, taken over North Africa, was featured in newspapers and magazines here and abroad as a "flying miracle." It shows a Boeing Flying Fortress en route home after a mid-air crash with a German Messerschmitt which almost cut the Boeing in two. The Fortress returned to its field and landed its crew safely. The German plane disintegrated in the collision. The Fortress was repaired and still is flying, according to word brought to Seattle by H.D. Ponceti, Boeing field service engineer, in North Africa and Italy.

(Original WWII Press Release for the photograph top right)

Battle Hardened B-17 Flying Fortress – A reputation of the B-17 was that it could take incredible punishment and return to base. Proving the point, the above left photograph shows the bomber *Dottie Jane* whose belly had been blasted out over Berlin by a direct hit. Even with significant damage and the loss of the Radio Operator who vanished into thin air, Pilot Arthur R. Socolofsky, from Chicago, IL, was able to bring the fortress home to its British base. The rest of *Dottie Jane's* crew survived. There are hundreds of similar stories and men that can still tell their tale. The photograph was released publically on March 30, 1944. (Both photographs by U.S. Army Air Forces)

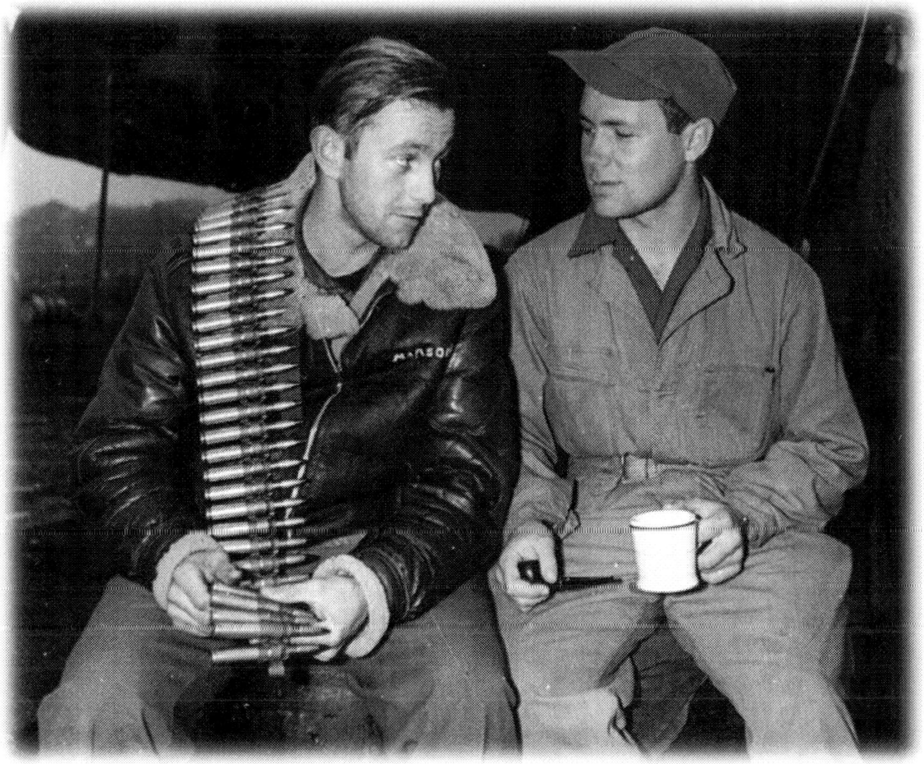

B-17 Flying Fortress Waist Gunner – Charles H. Marson (left) was photographed filling his ammunition belt and talking with a ground crew member. He was 23 and had already won the Air Force Medal. Before joining the U.S. Army Air Forces, he was a farmer in Boothbay Harbor, Maine where his wife and parents lived. The gentleman on the right is not identified. (British Combine)

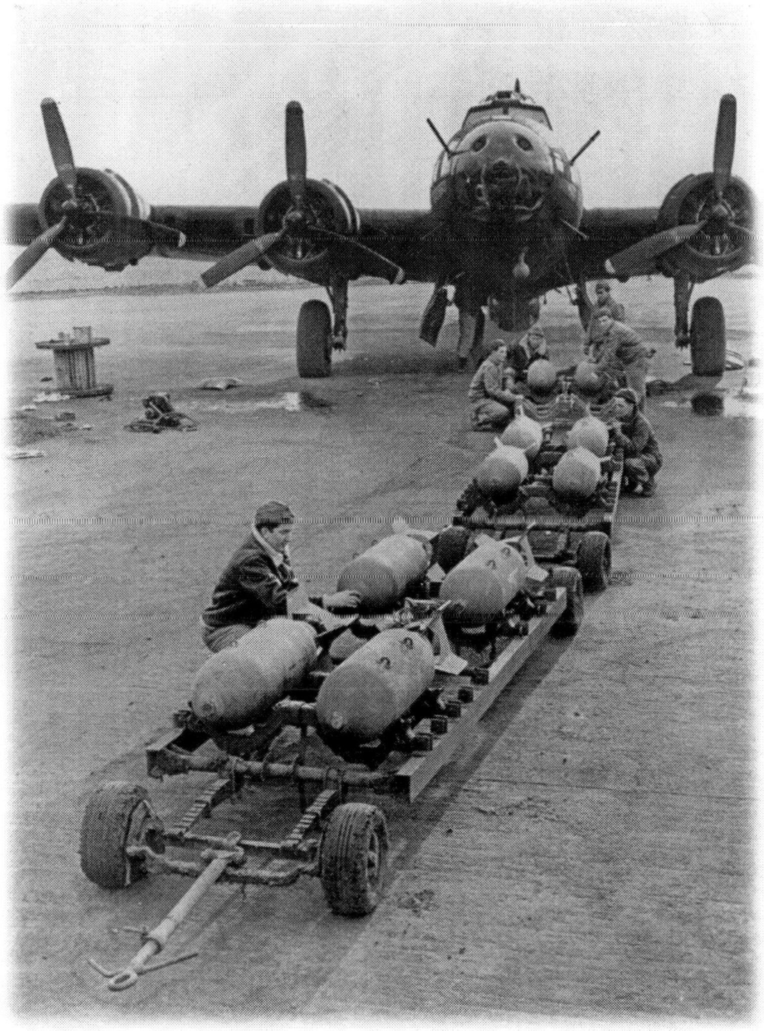

B-17 Flying Fortress – The business of 'bombing up', or loading the bomb bay of a B-17, took some effort. During a typical mission, a bomber carried between 4,500 and 8,000 pounds of bombs. Each bomb was loaded individually. In WWII, there were 640,000 tons loaded into the European Theatre B-17 bombers. The bombs were either hoisted by a hand crank or lifted by hand into position. (U.S. Air Forces)

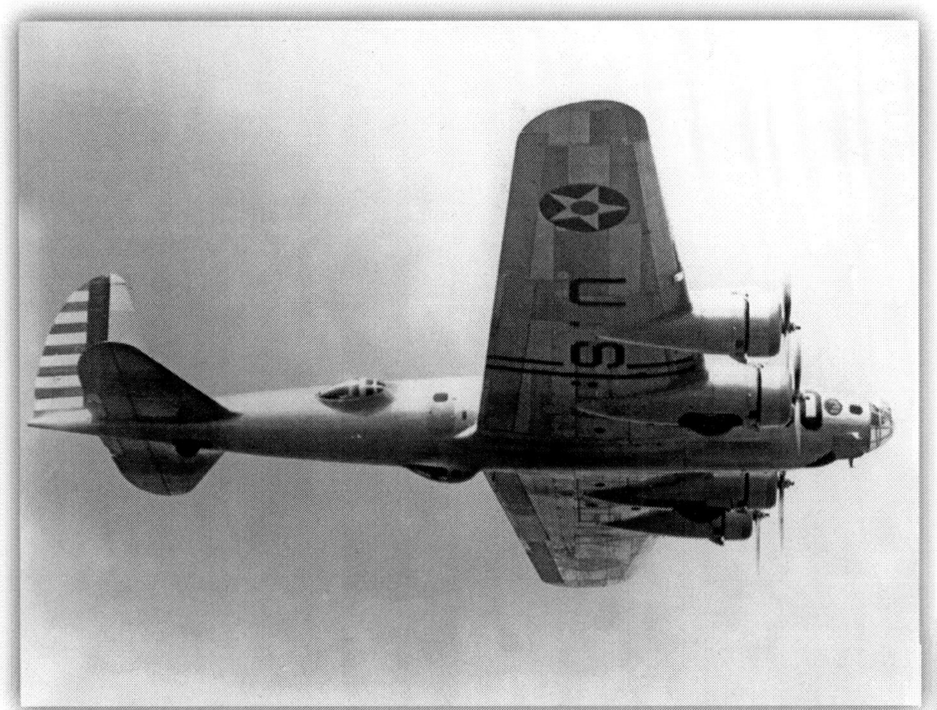

B-17A Flying Fortress

The earliest B-17, known as the B-17A, opened an entirely new field for bombardment and caused fighter plane design to follow suit. The bomber, although not considered battle ready, forced the strategy of air combat literally into the stratosphere. (U.S. Air Forces)

B-17G Flying Fortress

Battleship of the sky, the final B-17 production model, the B-17G, was produced in larger quantities than any previous model, 8,680. The aircraft defines "Flying Fortress" hosting thirteen .50-caliber machine guns – chin, top, ball and tail turrets; waist and cheek guns. With a service ceiling of more than 40,000 feet, the bomber could fly in excess of 300 mph and carry a maximum bomb load of nearly 10 ten tons. (U.S. Air Forces)

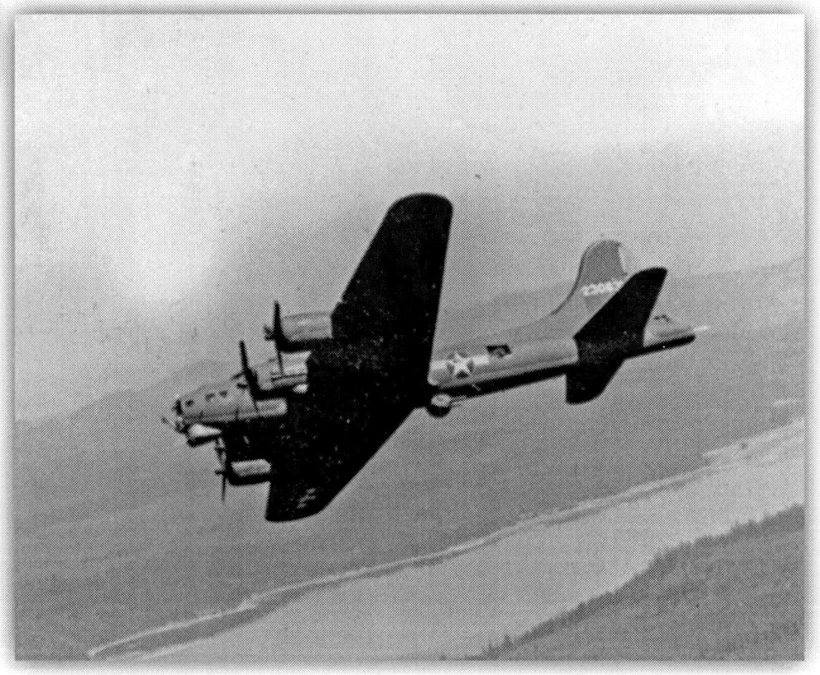

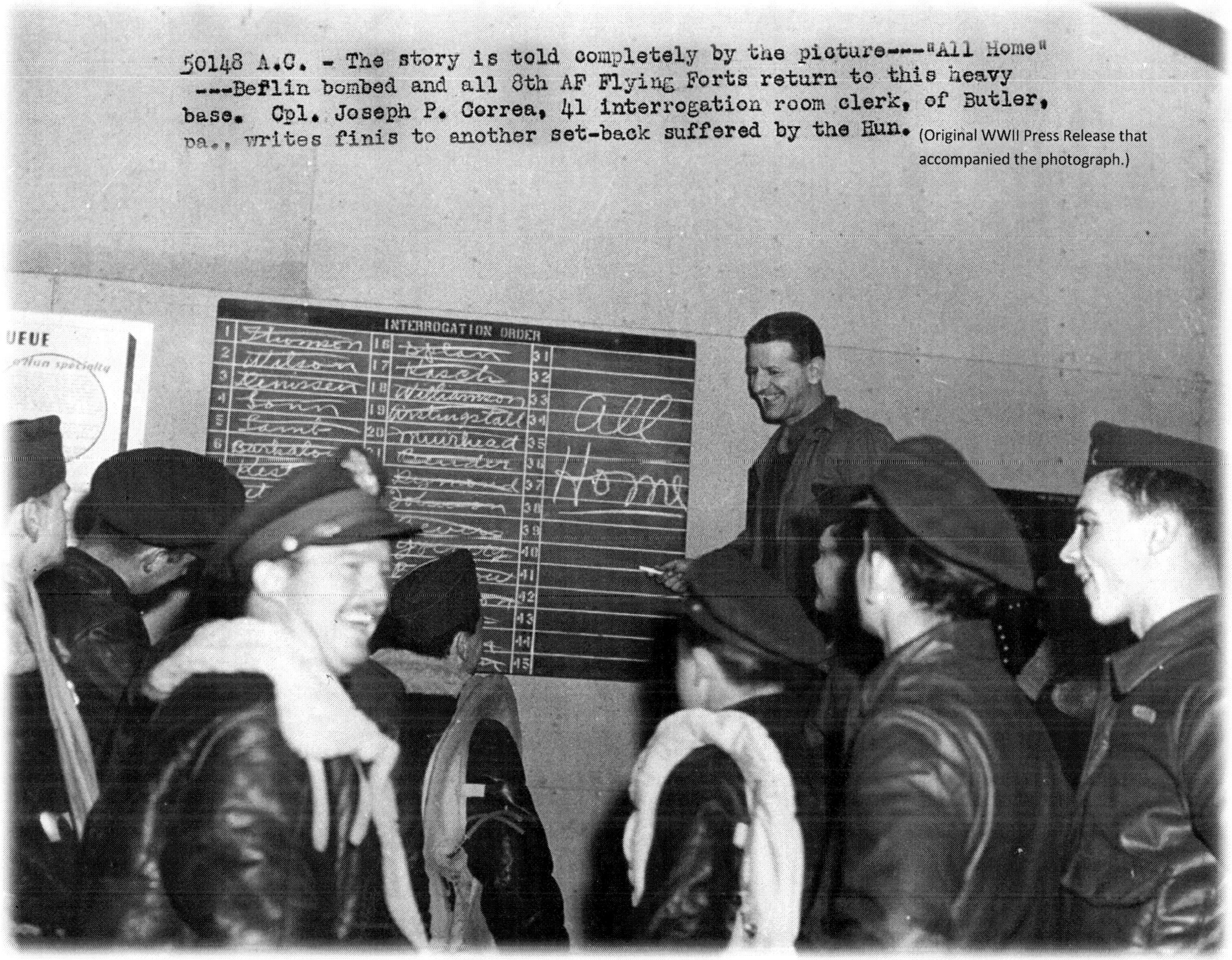

(Original WWII Press Release that accompanied the photograph.)

B-17 Flying Fortress – A total of 12,731 B-17s were built during WWII, 4,735 were lost in combat.[1] The bomber did its job but had aged during the war. At the conclusion of the war, the B-17 was quickly phased out of use as a bomber. Many were scrapped; some were used for VIP private service and some were used in testing new technology. (U.S. Air Forces) [1]www.b17.org/history/history.asp

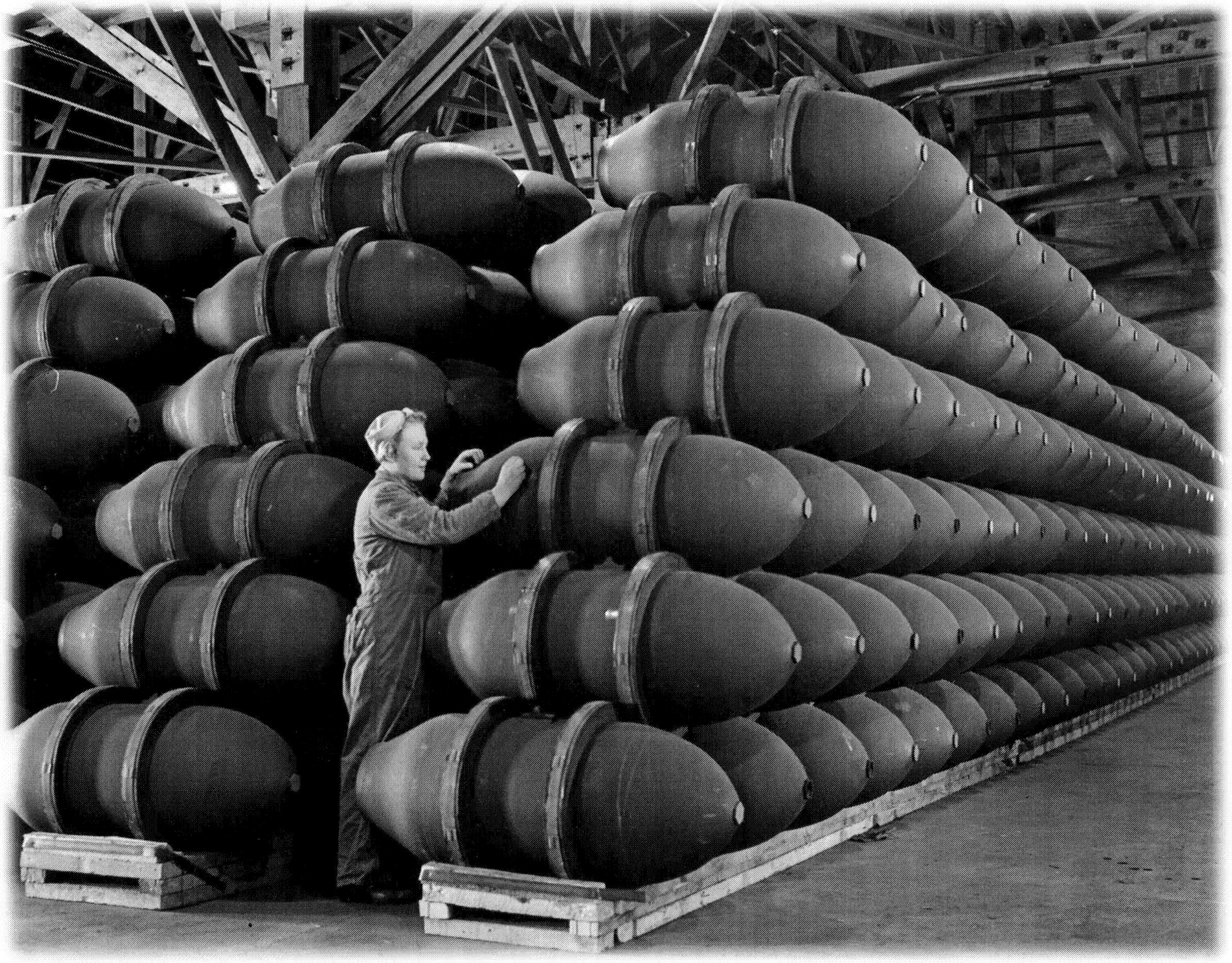

Bombs – A woman ordnance worker, known as a WOW during WWII, inspected a room of 1000-pound bomb casings. The casings were awaiting the installation of their deadly charge of explosives near Omaha, Nebraska. The war effort required warehouses of bomb casings because in Europe alone, 2,770,540 tons of bombs were dropped by the Allies.[1] (Unknown Photographer)

[1]The United States Strategic Bombing Survey, Statistical Appendix (European War), Feb 1947

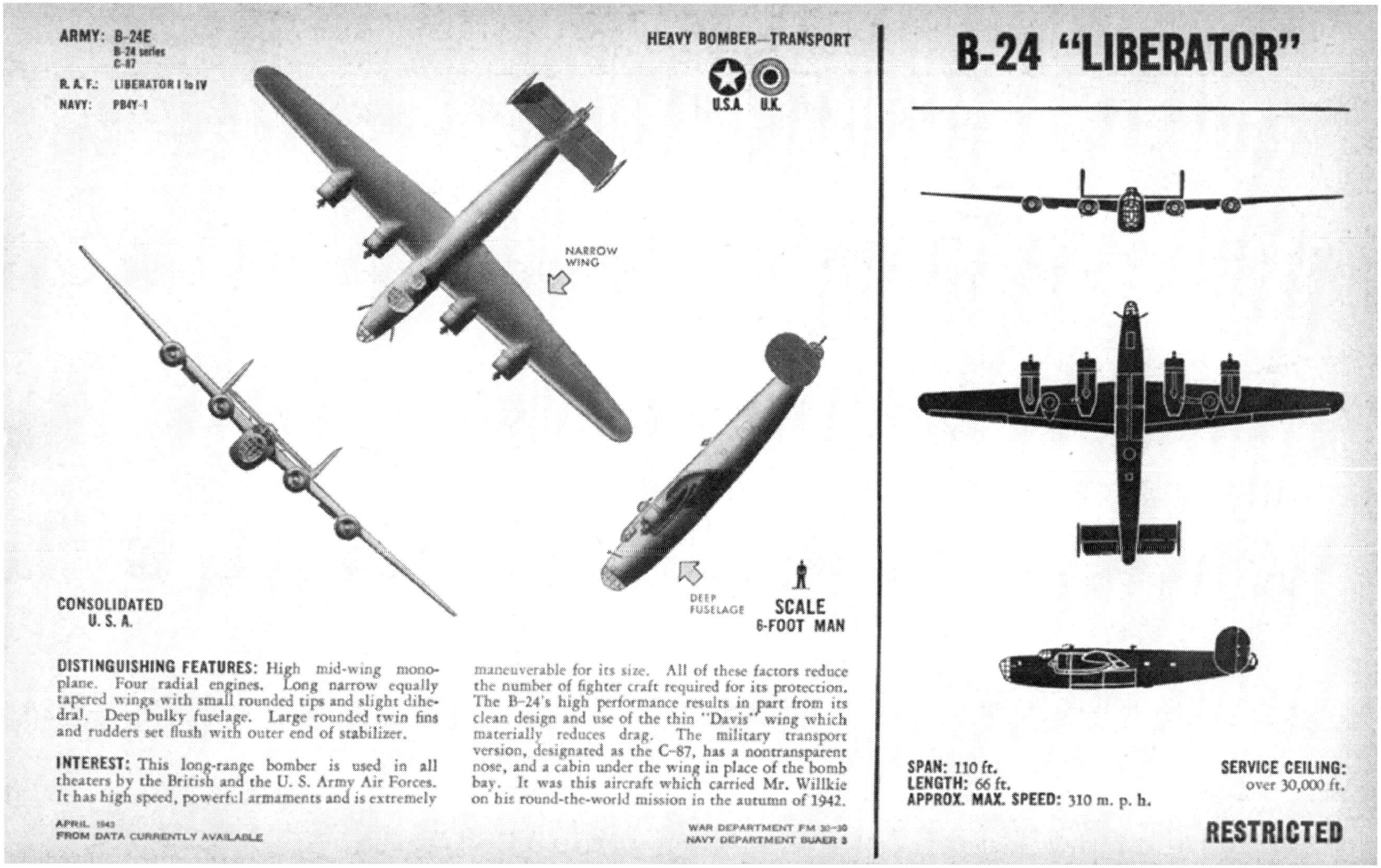

B-24 Liberator – The B-24, seen above in the Government's WWII recognition guide, clearly shows the aircraft's long, thin wings known as the "Davis" design. The B-24 was designed for fast manufacturing and had a higher top speed, greater range, and a heavier bomb load than the B-17. One drawback however was that it was more difficult to fly and many claimed it could not sustain the same battle damage. It was the first American bomber to use tricycle landing gear. Landon H. Brent remembered flying in the B-24, "Just before I finished 25 missions, they raised the number to 30 missions. Then when I finished 30 they gave me a three day pass to London, and told me to get ready to go home. That was a happy day. But when I got back from London I was told that they could find records of me flying only 29 missions and that I had to fly one more. That was a shock to put it mildly. I flew my 31st mission September 05, 1944 to bomb the Marshalling yards in Karlsruhe, Germany. We had a lot of damage from flak, but we made it back."[1] (War Department Recognition Guide 1943) [1]www.b24.net

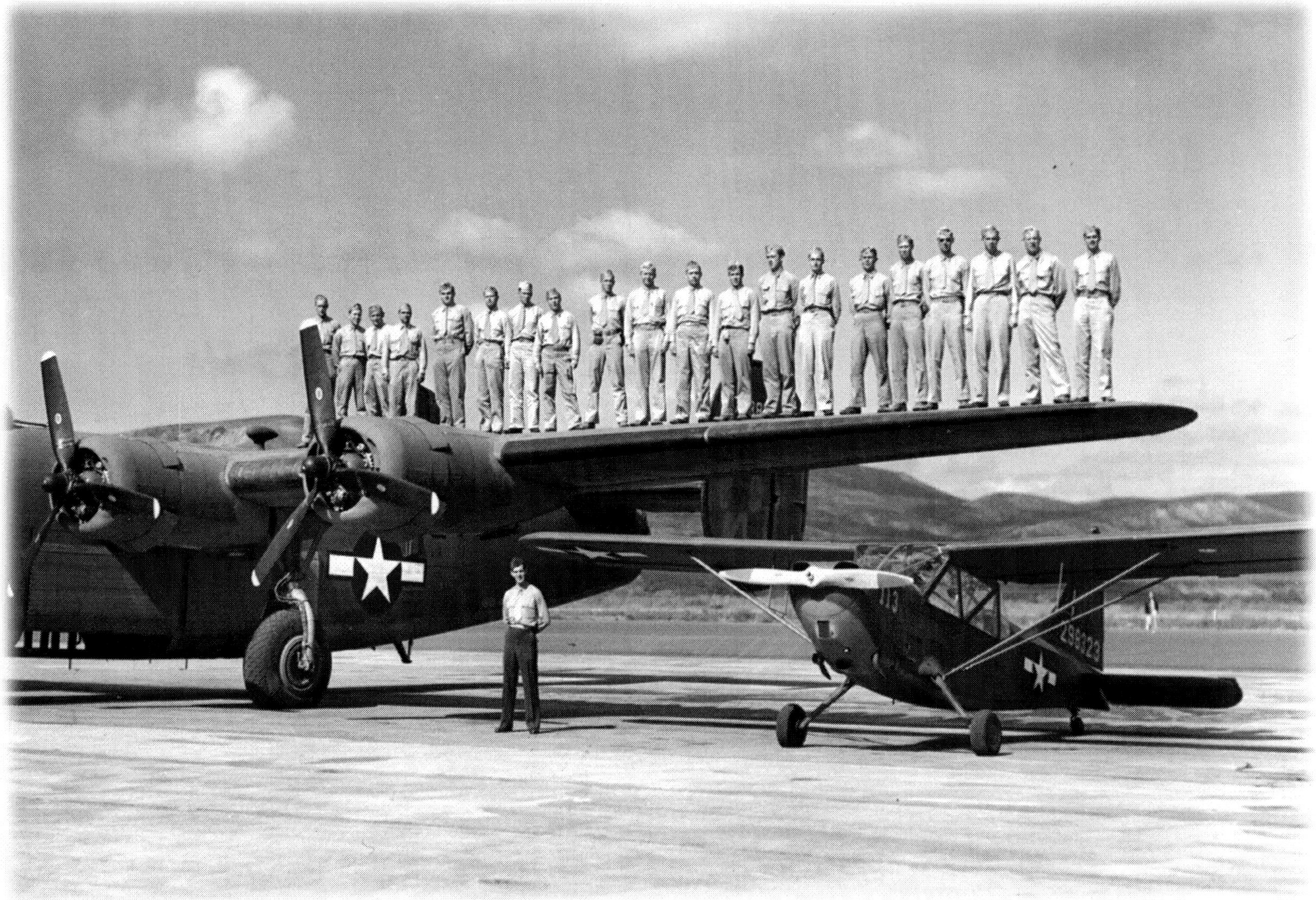

B-24 Liberator – The B-24 featured a new wing design called the Davis Wing. It was a high aspect-ratio wing where the ratio of the length to the width of the wing was quite slender. The design promised speed and greater range. To demonstrate its strength, Consolidated Vultee photographed a 56,000 pound B-24 with 20 men standing on the wing. Photographed alongside the bomber was a 2,100 pound flying Jeep (Stinson Sentinel) which was known to be a rugged aircraft. Many a merchantman sailing on Liberty Ships thought the Davis Wing's finest contributions to the war effort came when the wide-ranging Liberator could protect the last remaining open sea areas from prowling German U-boats. (Consolidated Vultee)

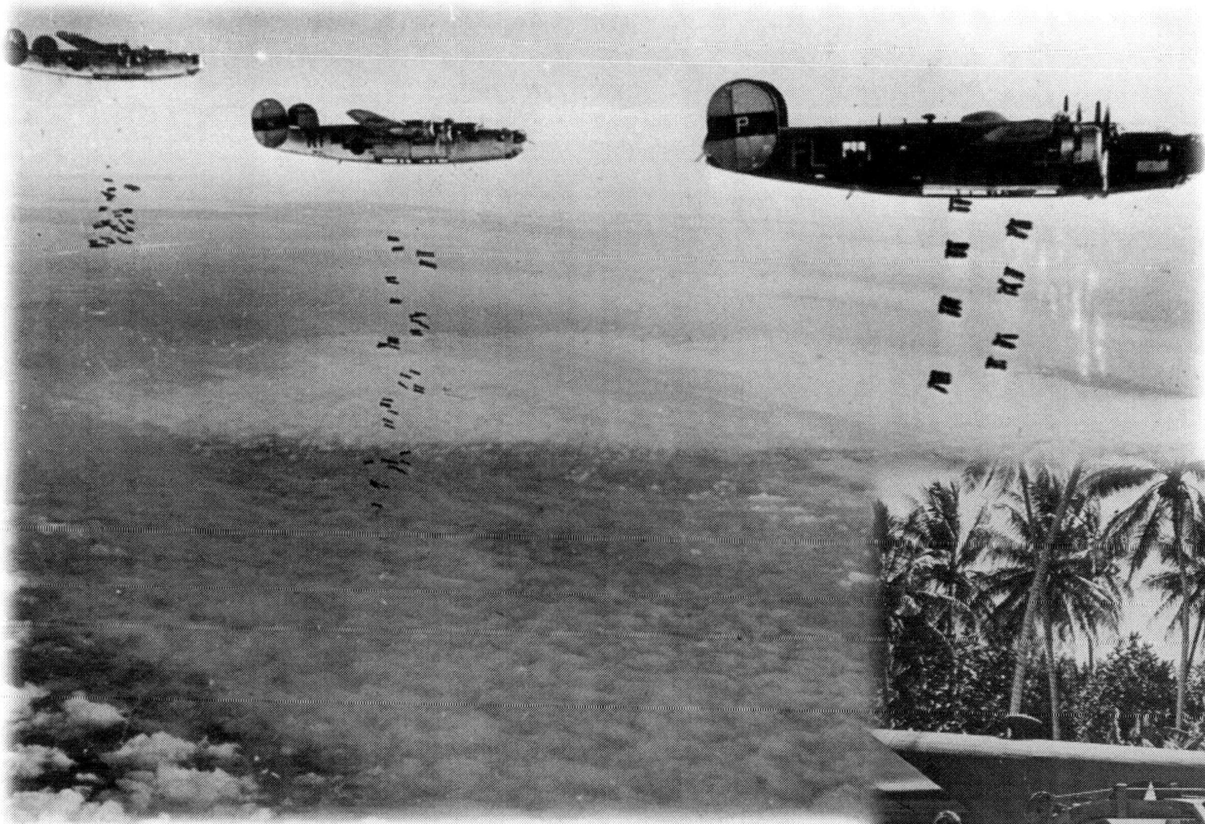

B-24 Liberator – These 8th Air Force B-24s were photographed on March 9, 1945 as they were unloading their bombs on railroad marshalling yards at Rheine, Germany. Keeping a close, tight formation was the secret of survival from fighter plane attacks because the concentrated firepower of several bombers made diving into the formation a risky maneuver. The pilot of *Pregnant Peg*, a B-24 bomber, said, "We had trouble on the 29th, right from the start because of weather conditions. The initial head on tactics of the fighters spread us out a little too. The B-24 wasn't as easy to fly as the more famous B-17, but it was, in my opinion, the better of the two aircraft. It took more muscle to maneuver in formation, but it was faster, carried a bigger bomb load, and I liked the Pratt and Whitney engines."[1] (Army Air Force)

[1]www.b24.net

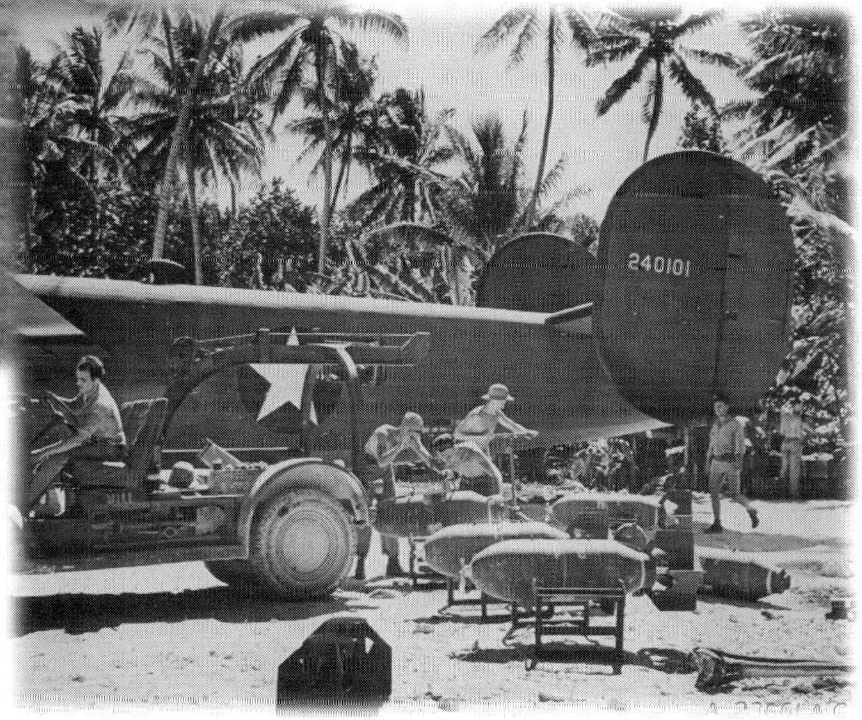

B-24 Liberator – Ground crews were photographed loading bombs into a B-24 that had landed on the Pacific Island, Tarawa, in the Northern Gilbert Islands. The airbase had been an enemy airbase until it was captured by the U.S. Marine Corps in late November 1943. (U.S. Army Air Forces)

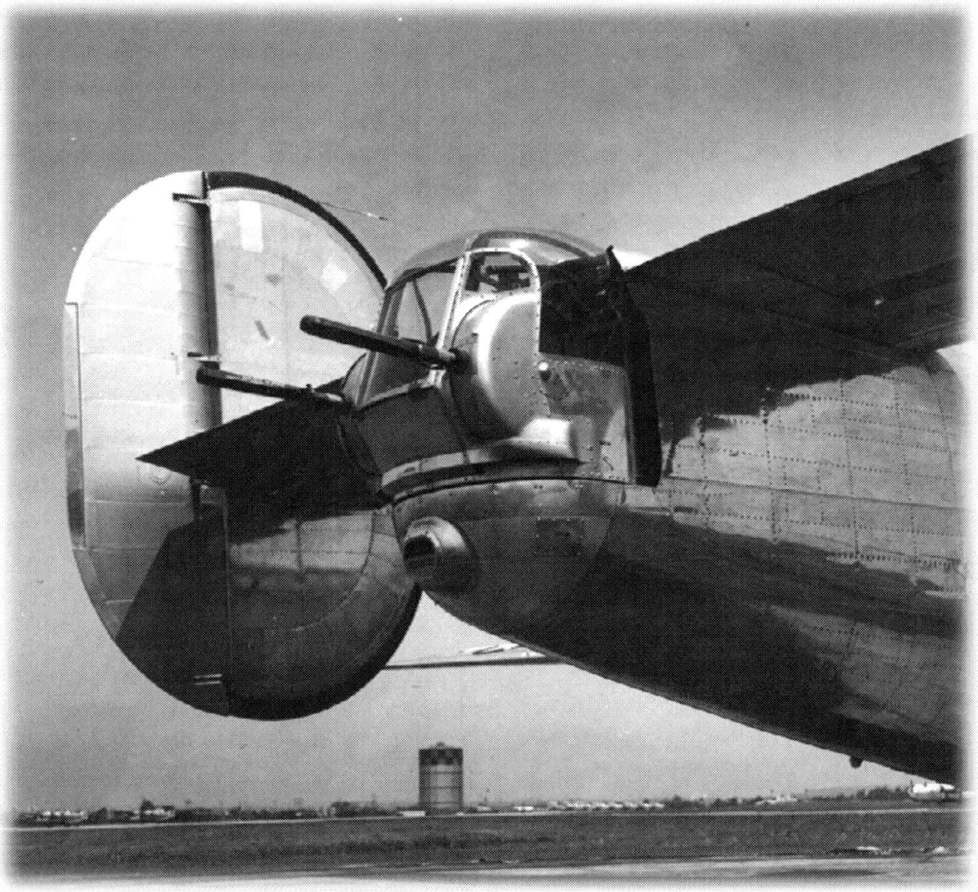

B-24 Liberator – The tail gunner could be considered the most important defensive position aboard a Liberator. The gunner manned a pair of .50 caliber heavy machine guns housed in a powered rotating gun turret. He was charged with defense of the aircraft's "six" position which encountered trailing enemy fighters eager to attack. (Rudy Arnold)

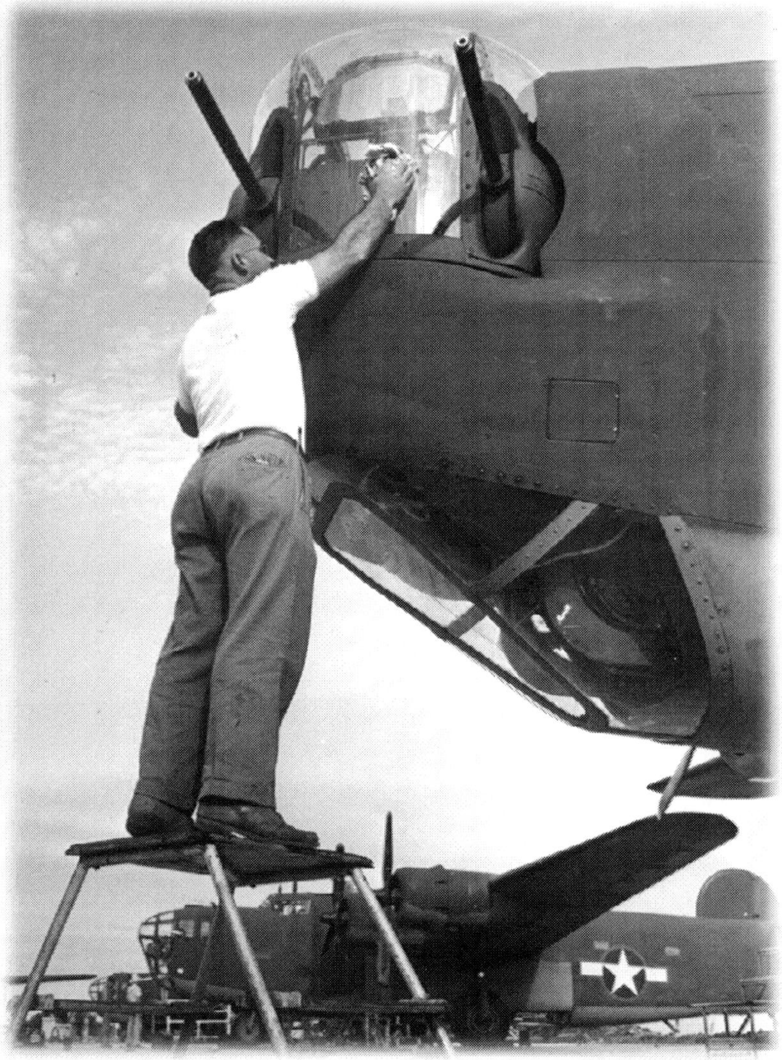

B-24 Liberator – On the right, a worker cleaned the new electrically operated nose turret. The B-24 was produced in several locations. Consolidated Aircraft used its plant in San Diego and built a large new plant outside Fort Worth, Texas. Douglas Aircraft built B-24s in Tulsa, Oklahoma. North American Aviation built them in Dallas, Texas. The Ford Motor Company built a vast new factory at Willow Run, Michigan that dwarfed other locations. In all, Ford produced half of the 18,000 B-24s built.[1] (Consolidated Vultee) [1]Nolan, Jenny (January 28, 1997). Willow Run and the Arsenal of Democracy. Michigan History, *The Detroit News*.

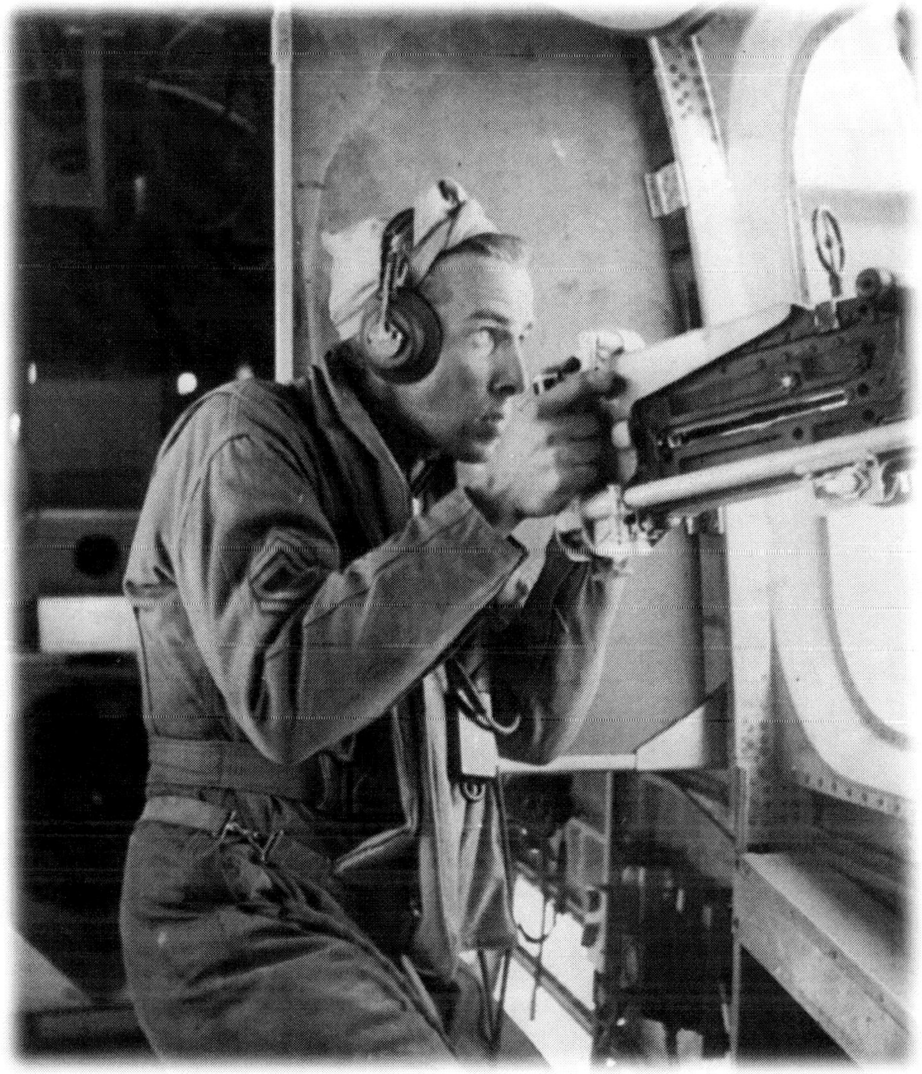

B-24 Liberator Waist Gunner

Technical Sergeant Vincent F. Wachs of Pittston, PA, left, was photographed firing fast action .50 caliber shells to the enemy above the Mediterranean Sea. His formation of B-24s was en-route from a Libyan Airport to bomb Grottagli, Italy when they were jumped by 12 German Messerschmitt 109s in June 1943. (ACME)

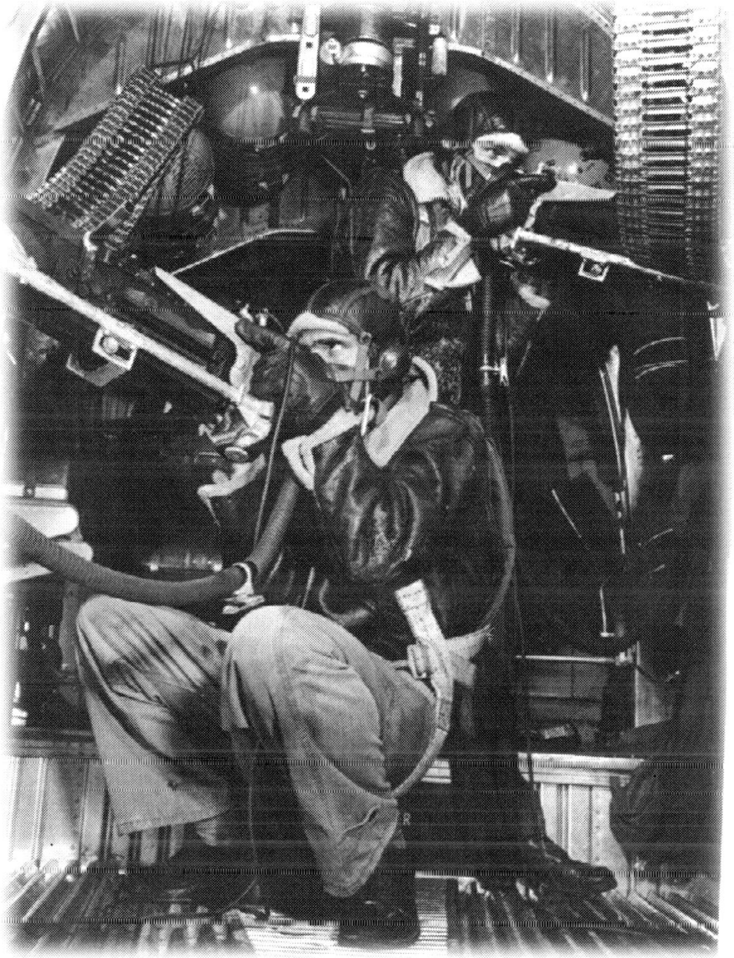

B-24 Liberator Waist Gunners

In the photograph to the right, Private George R. Folds of Evanston, IL, left, and Private James Boulware of Charleston, SC, were photographed at their stations wearing oxygen masks and fleece-lined jackets while flying at 20,000 feet. The gunners were at the Army Air Force Training Command in Harlingen, Texas. (Army Air Force)

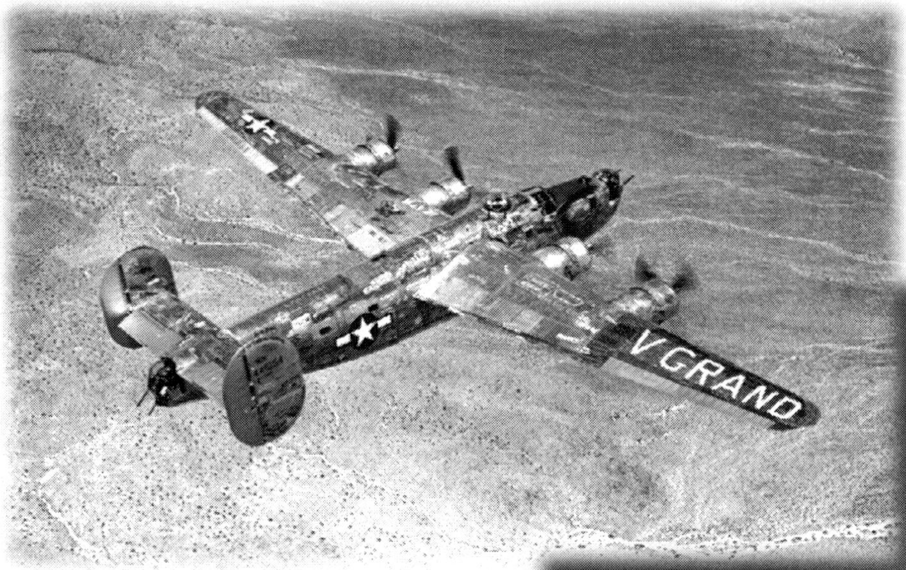

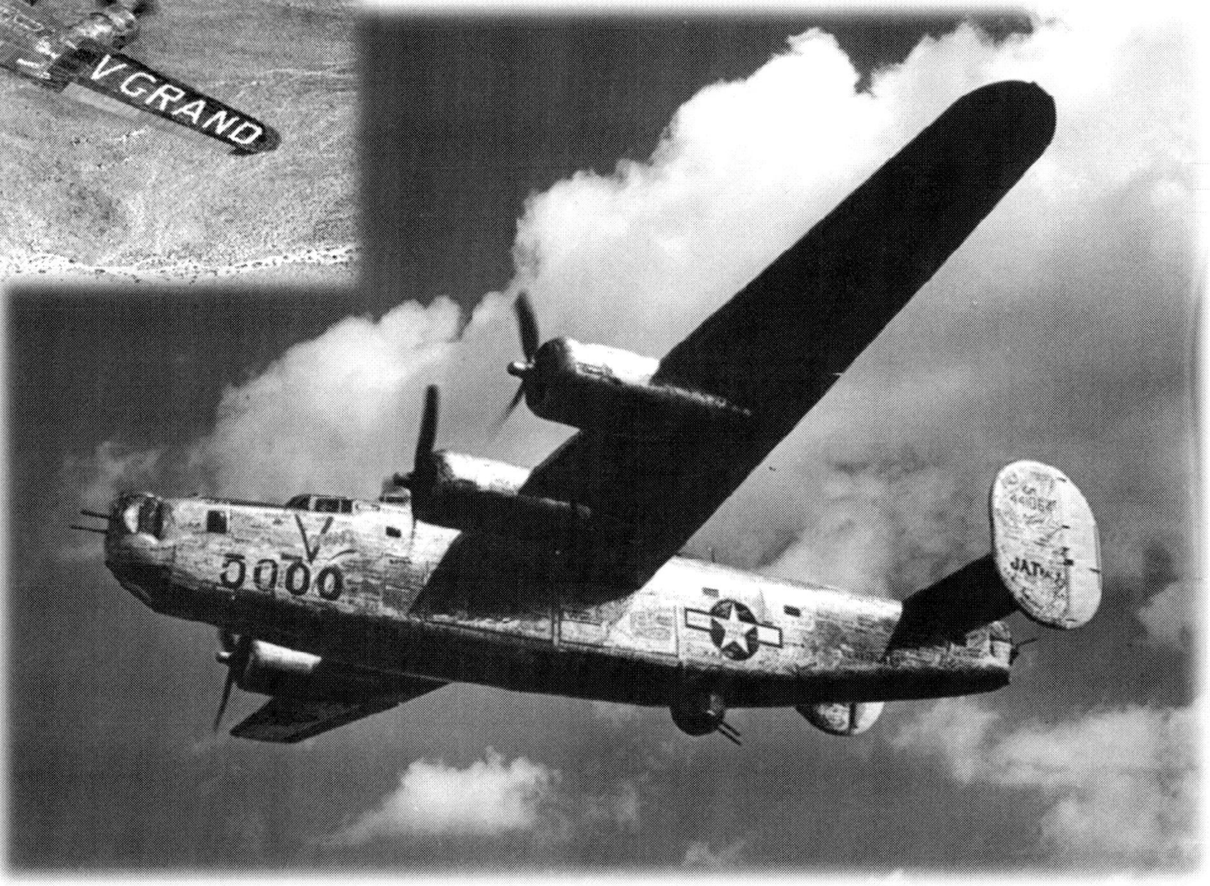

B-24 Liberator – Looks can be deceiving because this B-24 only looks like it is the 5,000th bomber. It's actually the 7,000th bomber to be built but the 5,000th to roll off the assembly lines at the Consolidated Vultee Aircraft plant in San Diego. The aircraft was decorated with the signatures of 7,000 employees prior to its 10 man crew flying it into action. The pilot this day was 2nd Lieutenant James D. Ayers of Hollywood, CA. The publicity surrounding high production numbers of Allied aircraft shocked our enemies who questioned how America could produce so many aircraft. (Both photographs by 1st Air Force - Hugelmeyer)

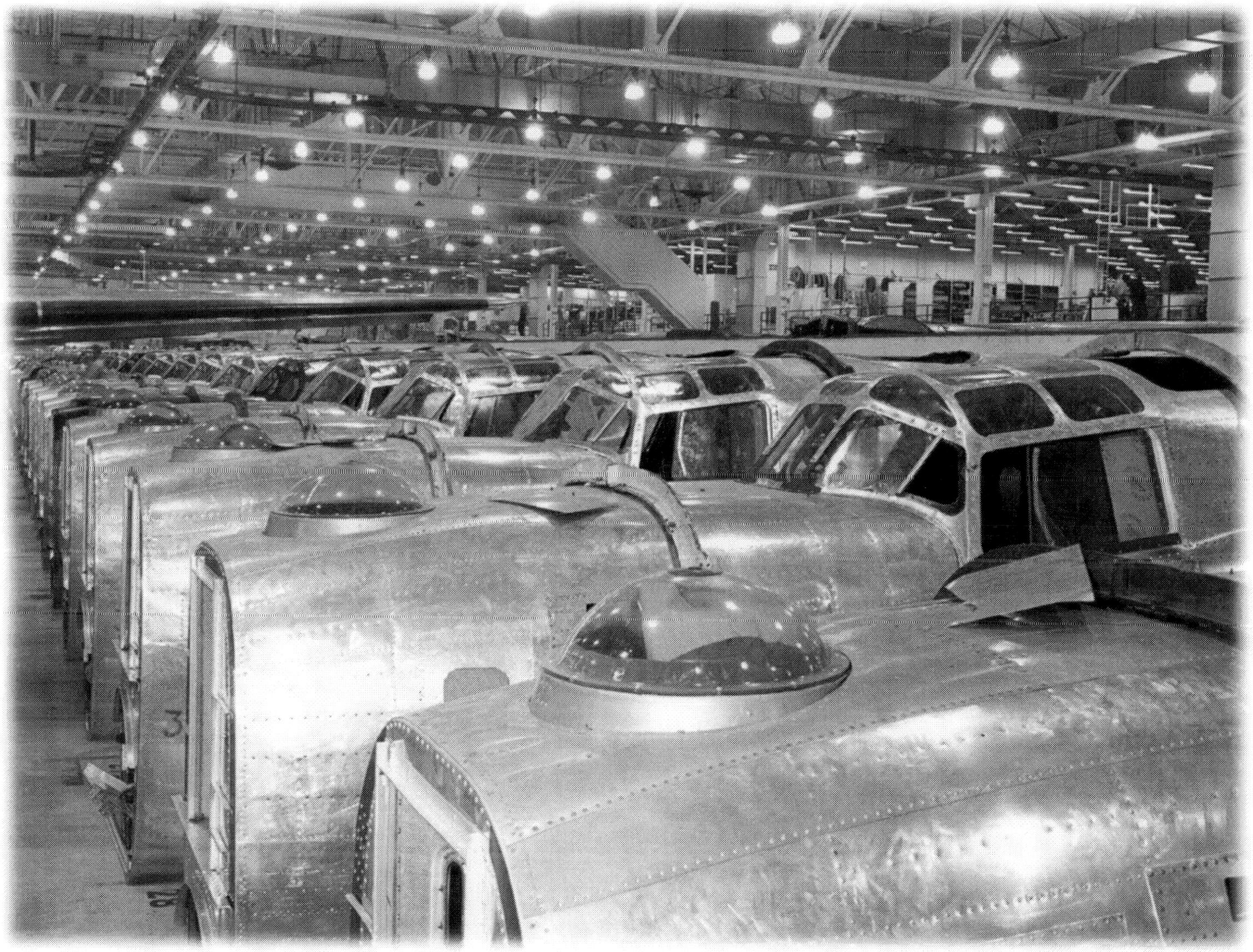

B-24 Liberator – Pictured above is a view of the Nose Line for Liberator construction at the North American Aviation plant in Texas. The facility produced B-24G and B-24J models until mid-1944 when planners determined that the San Diego and Willow Run plants could maintain all necessary B-24 production. Both the Douglas and North American plants ceased Liberator production. Liberators dropped over 630,000 tons of bombs and destroyed several thousand enemy aircraft during WWII. With a range of 2,200 miles, the aircraft proved useful for many tasks. A few were converted to carry the first U.S. air-to-surface radar-guided missile, the Bat, which sank a Japanese naval destroyer in April 1945.[1] (North American Aviation, Inc.) [1]www.aviation-history.com

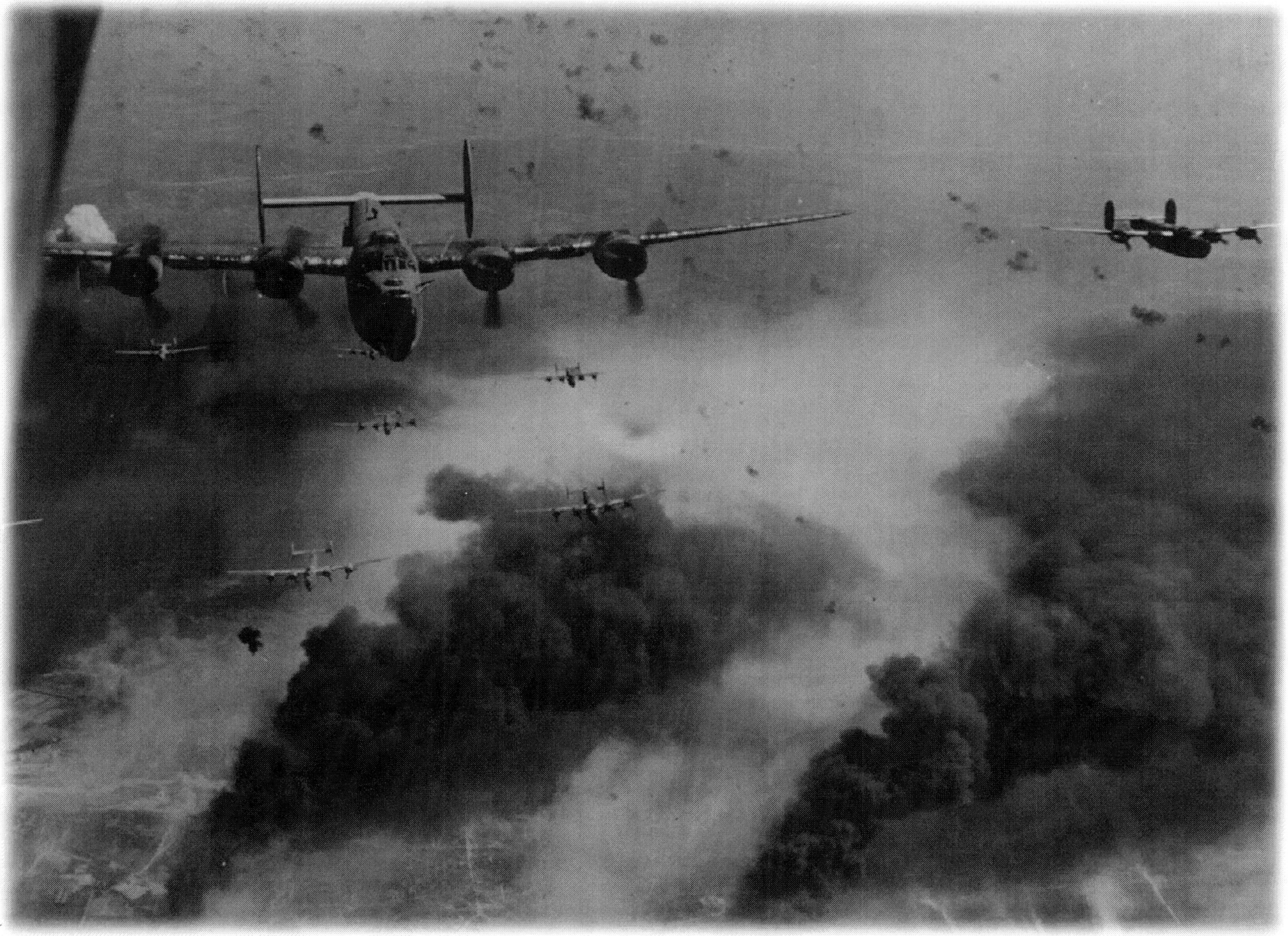

B-24 Liberator – A photograph from a 15th Air Force B-24 attack on the Ploesti, Rumania oil refineries captured the danger and randomness of battle. Inspecting the photograph closely shows the tell-tale puffs of flak smoke over the target, the destruction below and fortunately no bombers trailing smoke from a damaged engine. Closer inspection shows all of the props are turning giving these bombers a better chance of getting home. When they made it back to base it was one more mission in the log book. (U.S. Air Forces)

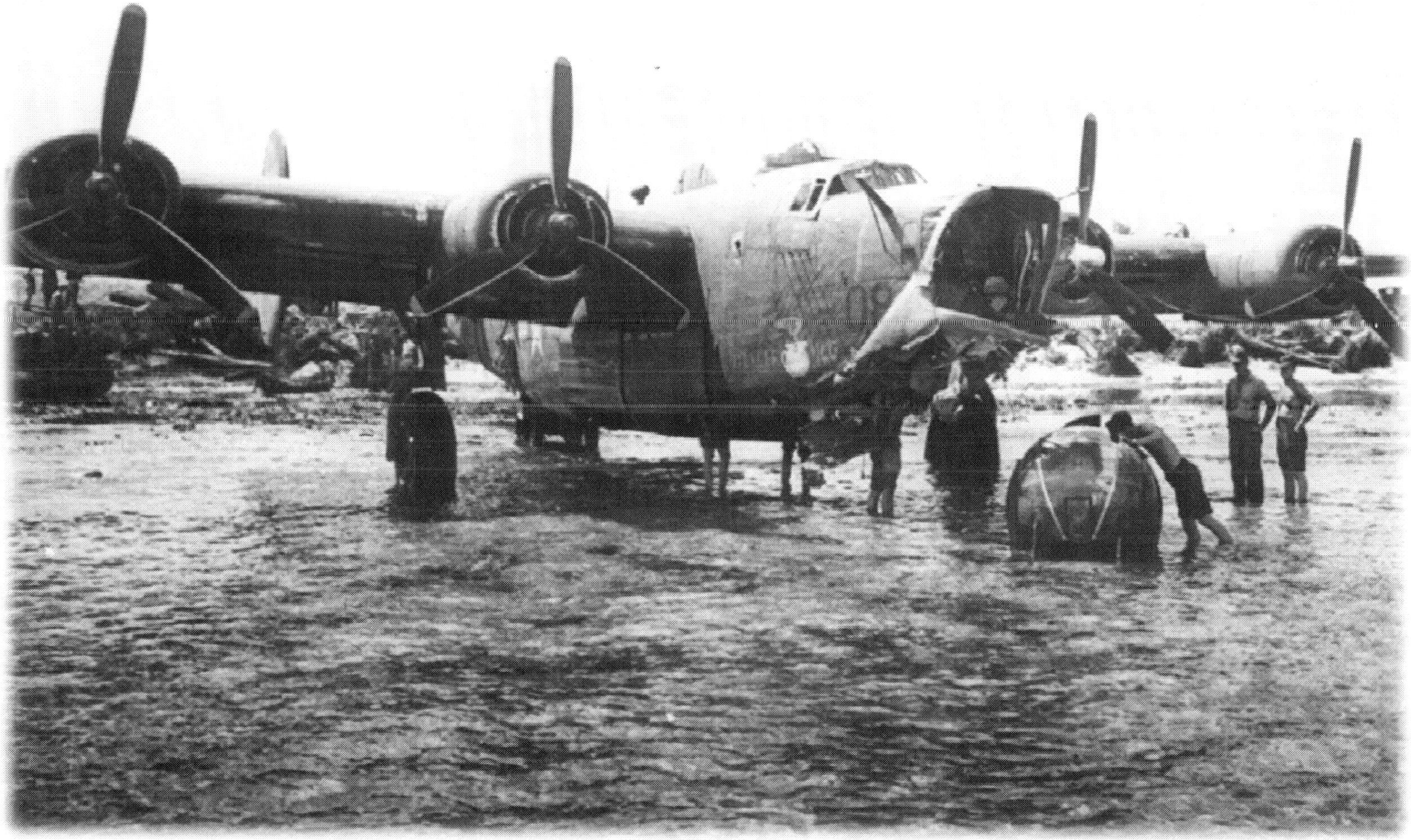

U.S. Navy Liberator PB4Y-1 – In September 1944, this PB4Y-1, *Thunder Mug*, had its hydraulic braking system destroyed by Japanese flak during a raid on Puluwat Atoll in the Caroline Islands. Trying to land, the Liberator ran off the landing strip and nosed into the ocean. Although sad to lose an aircraft, the crew survived. The Allies certainly enjoyed a return on investment with *Thunder Mug*. It was the aircraft that Commander Norman M. Miller, U.S. Navy, used to make the majority of his astonishing tally of 66 Japanese ships sunk or damaged. (U.S. Navy)

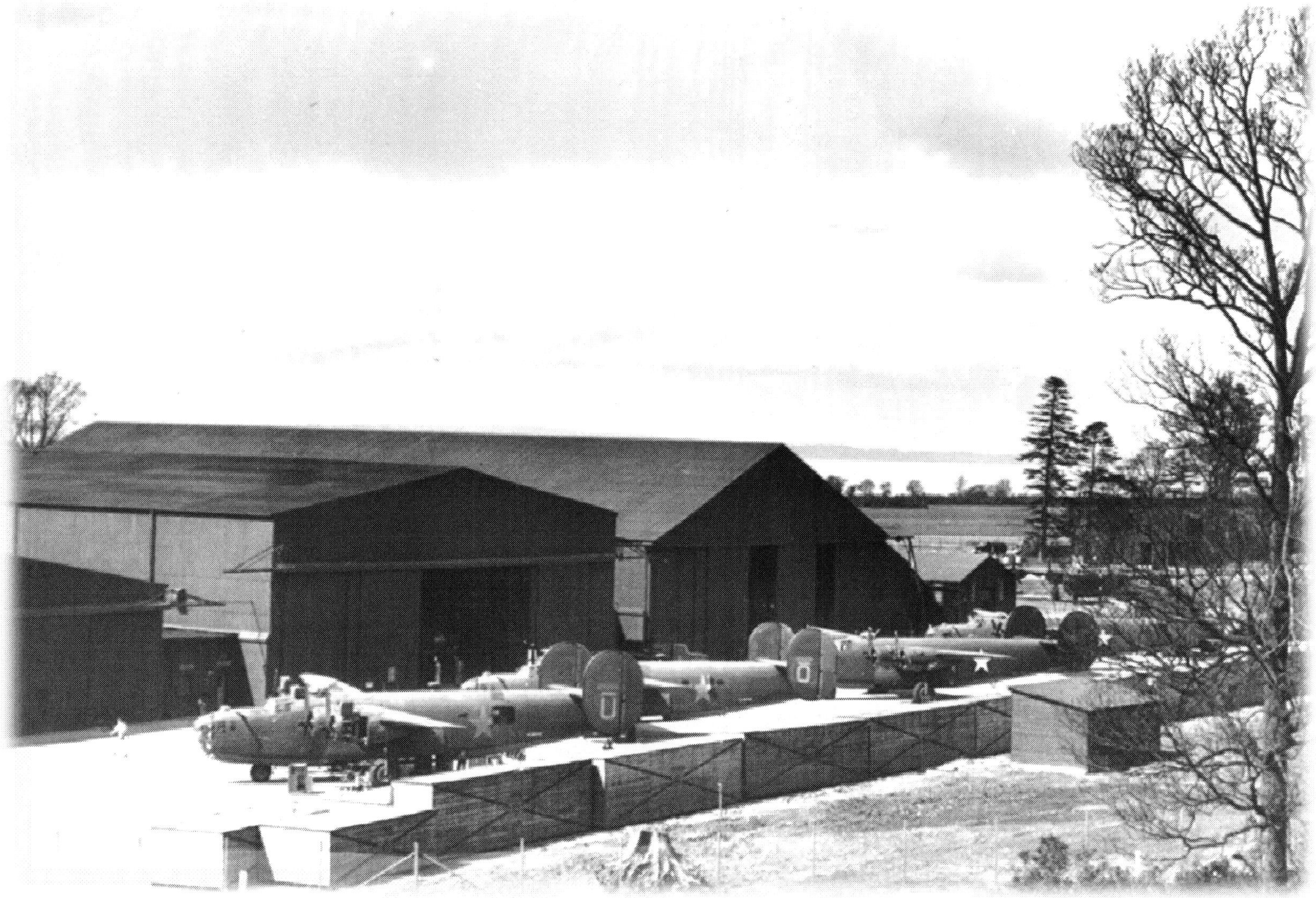

B-24 Liberator – Northern Ireland served as a service center for all types of American aircraft. Seen above, B-24s are lined up in front of three of the 17 hangars used by the Lockheed Overseas Corporation for the repair, service and modification of American-made aircraft. In the forefront of the photograph is a row of wooden crates that contain disassembled Brewster Bermuda dive bombers. They were ready for assembly and soon would be flown into combat by the British. (Lockheed Aircraft Corporation)

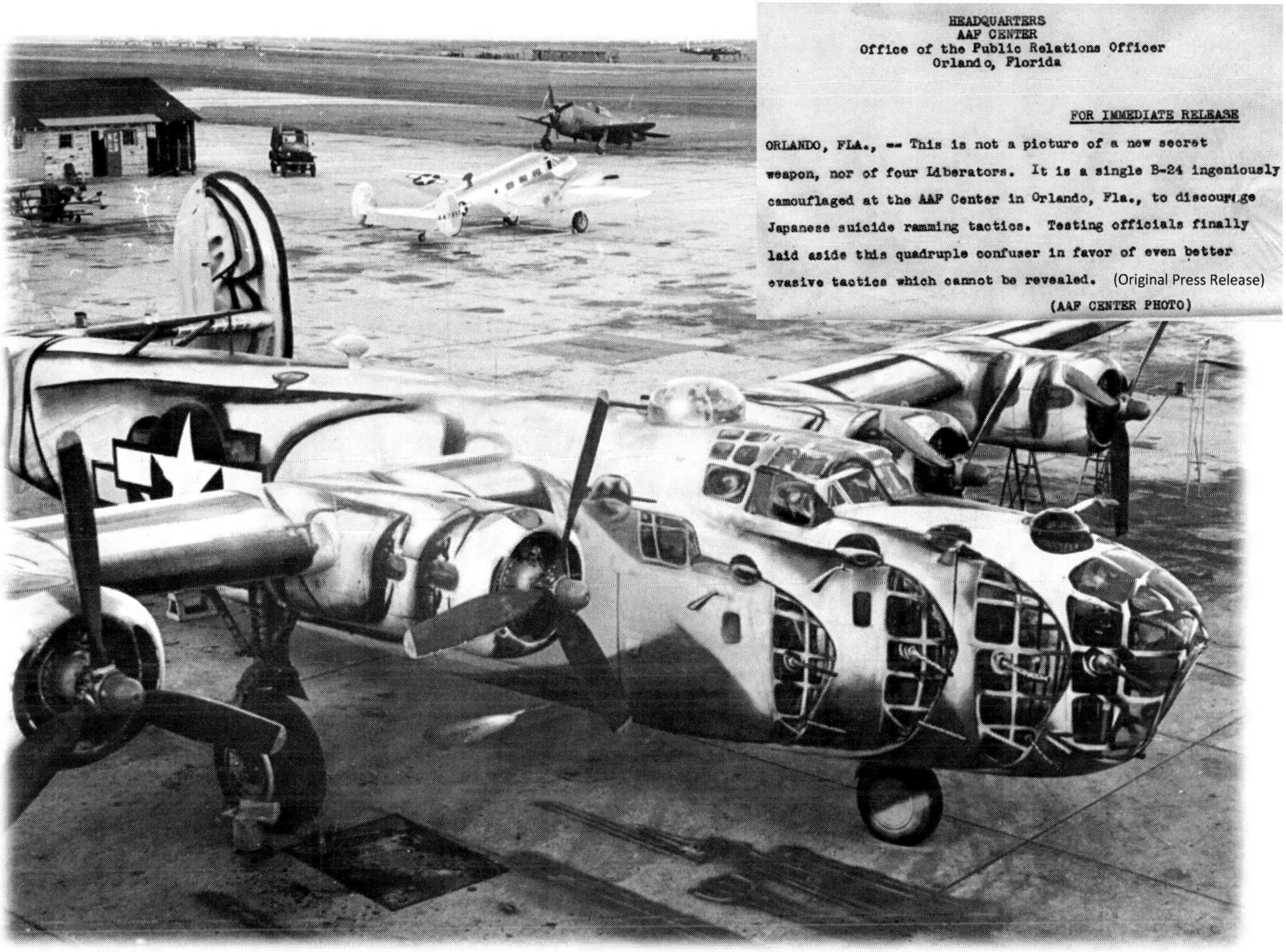

HEADQUARTERS
AAF CENTER
Office of the Public Relations Officer
Orlando, Florida

FOR IMMEDIATE RELEASE

ORLANDO, FLA., -- This is not a picture of a new secret weapon, nor of four Liberators. It is a single B-24 ingeniously camouflaged at the AAF Center in Orlando, Fla., to discourage Japanese suicide ramming tactics. Testing officials finally laid aside this quadruple confuser in favor of even better evasive tactics which cannot be revealed. (Original Press Release)

(AAF CENTER PHOTO)

B-24 Liberator – This B-24 was painted in a design being considered to discourage enemy pilots from trying to ram the aircraft in flight. In the Far East, fighting the Japanese, Allied aircraft suffered from attacks of suicide pilots similar to the Navy's experience with Kamikaze. Many different techniques were deployed to avoid the suicide flyers. (Army Air Forces)

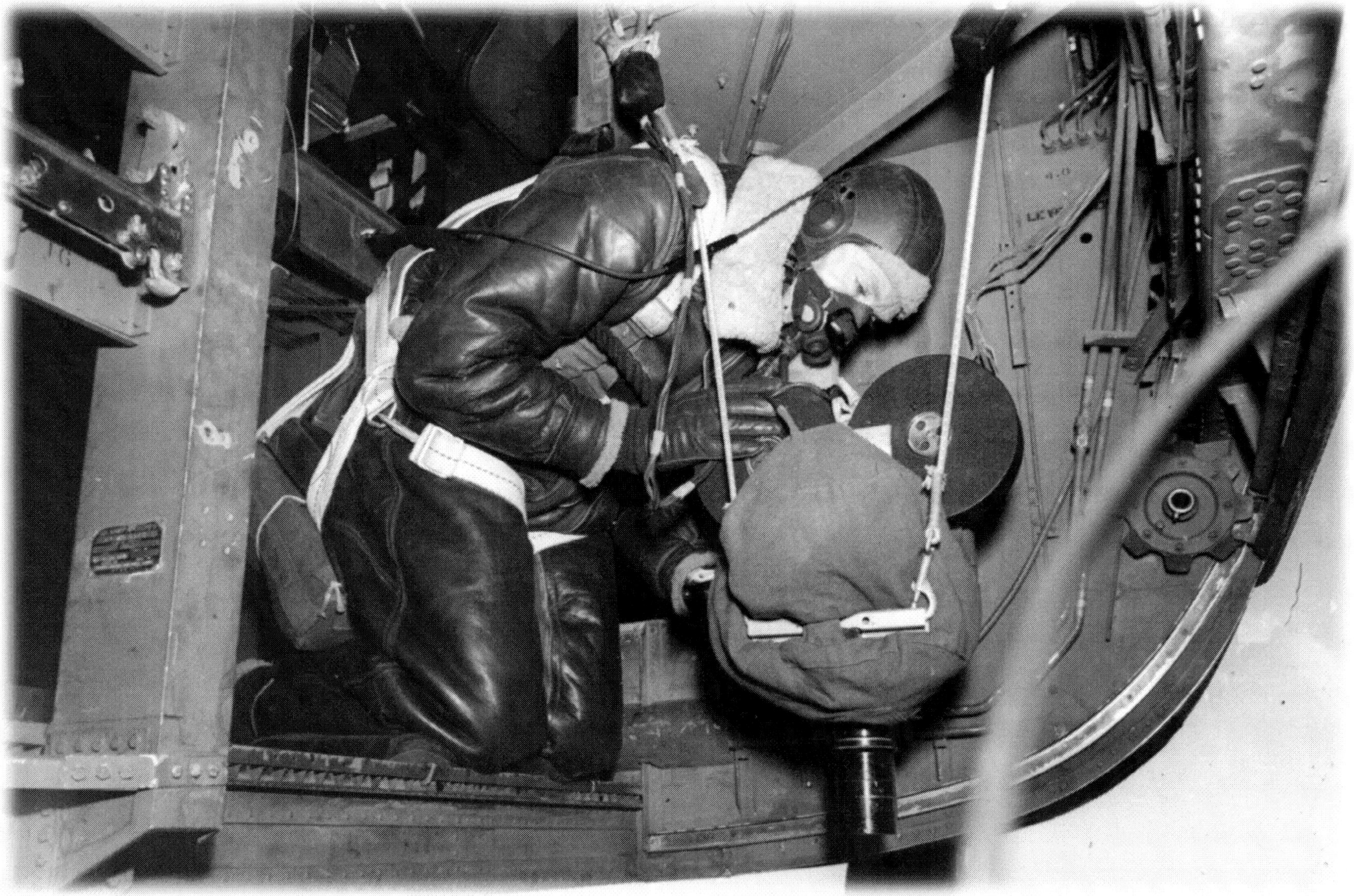

B-24 Liberator – Obtaining a photograph of enemy positions or battle damage from a mission was not always as easy as scheduling a formal aerial recognizance aircraft. John D. Craig demonstrated how he operated a Type A-3 Aircraft camera with his own specially designed B-24 heated camera mounting jig. His technique, shooting from the catwalk of a B-24 Liberator, simplified the procedures for high altitude target photography. From the looks of this photograph, it simplified the process for men who didn't mind heights and leaning over an open bomb bay tens of thousands of feet above the earth. (War Department)

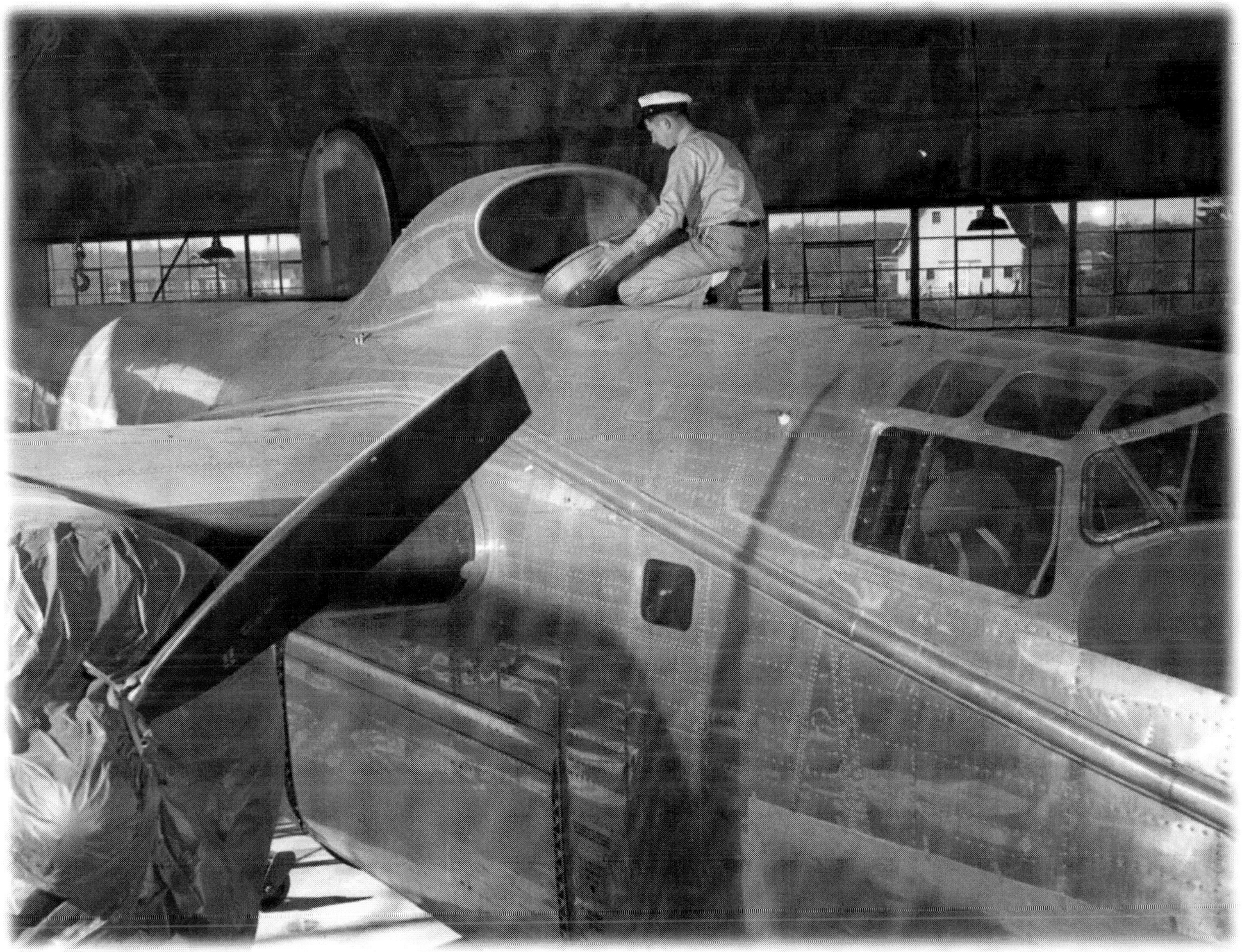

B-24 Liberator – This B-24 Liberator was fitted with an experimental air scoop designed to feed air to a turbine inside the aircraft's fuselage. The aircraft was converted by General Electric as part of their flying laboratory. The G.E. Flight Test Center was located at the Schenectady County Airport in New York. (General Electric)

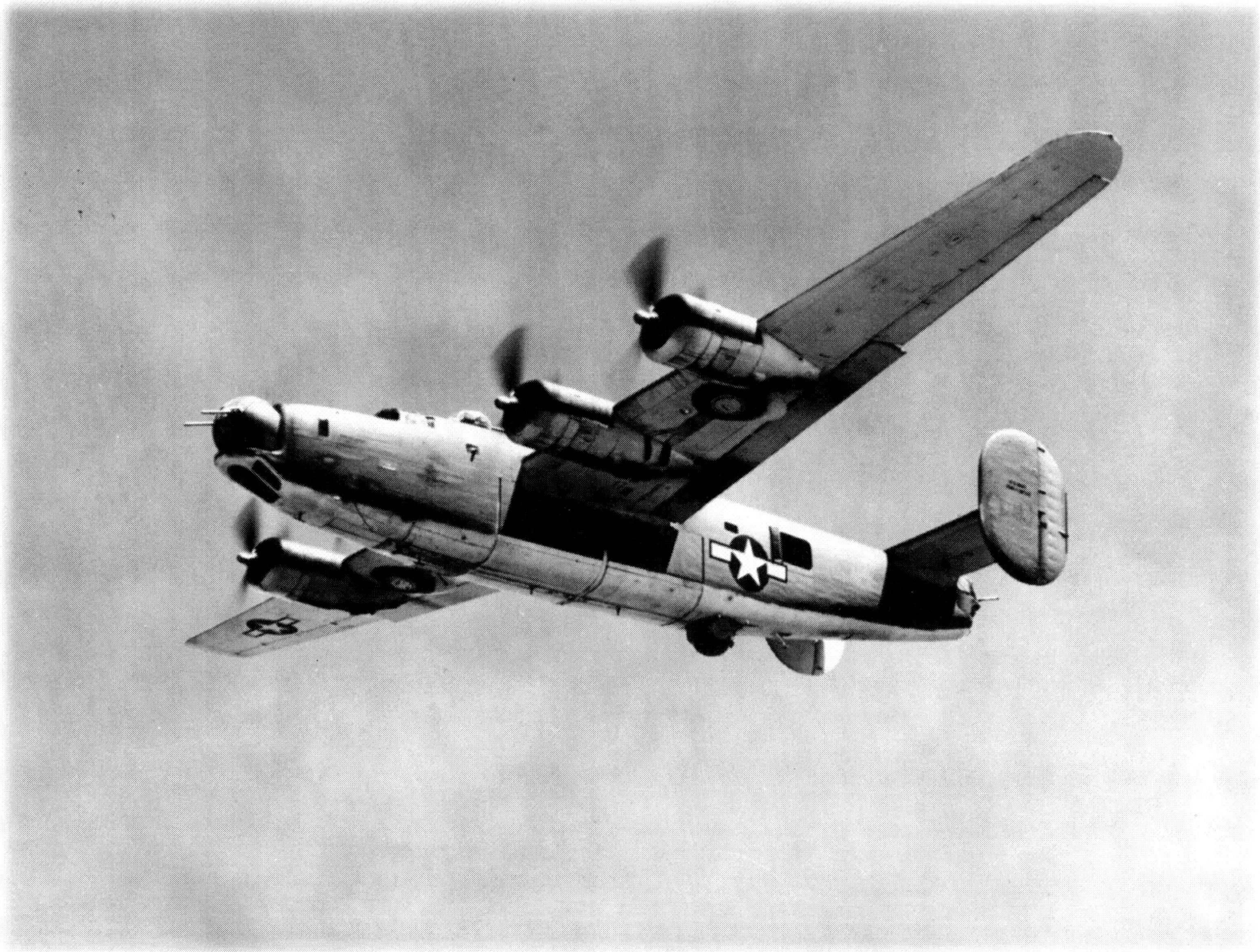

PB4Y-1 Navy – The aircraft above was the Navy's version of the B-24 designated the PB4Y-1. It was manufactured at the Consolidated Vultee plant in San Diego and except for a few modifications which made it more suitable for naval bombing and patrol missions, the PB4Y-1 was the same as the Army's B-24. The Navy PB4Y-1 featured a blue-and-white camouflage and a ball turret in its nose which distinguished it from Army aircraft. (Consolidated Vultee)

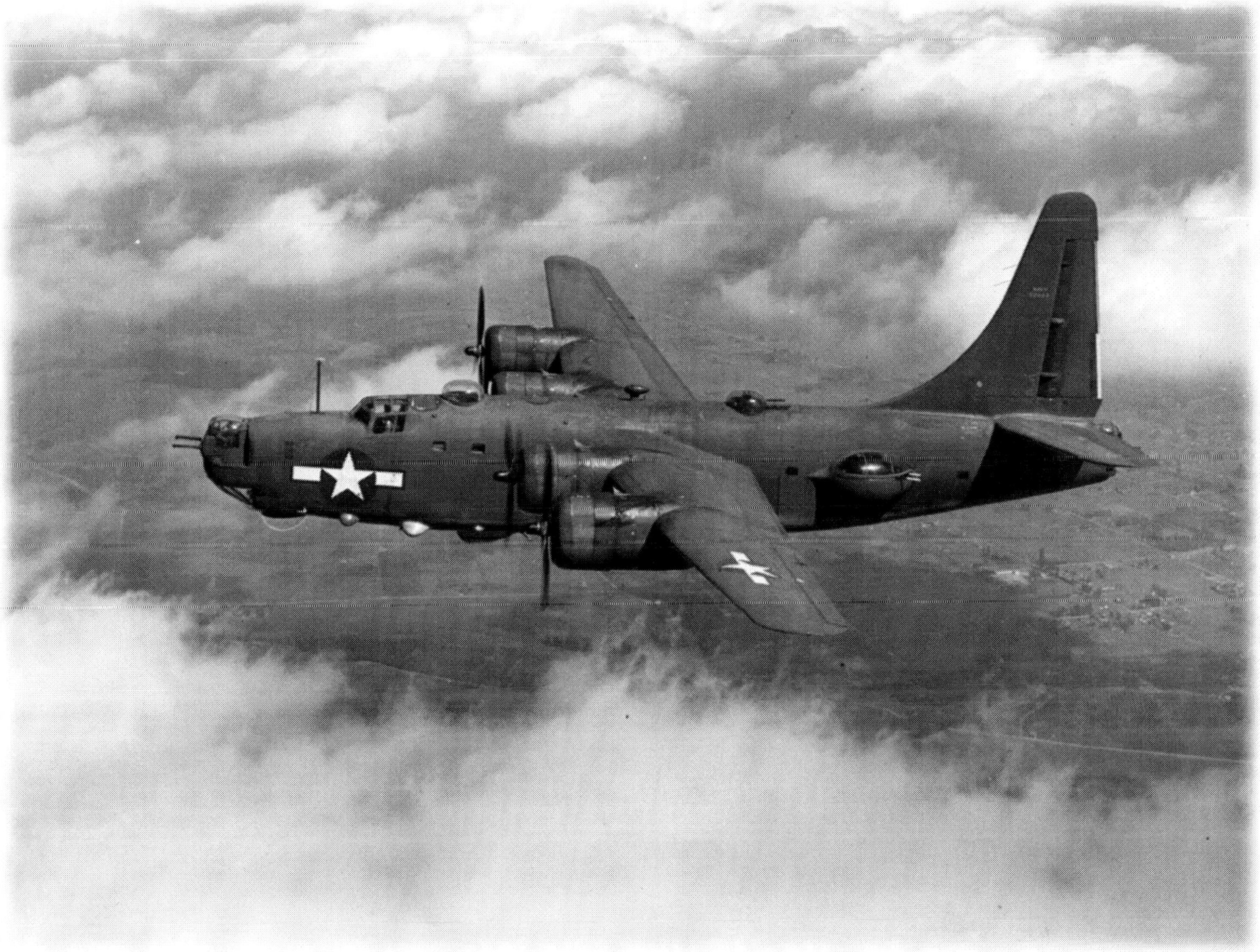

PB4Y-2 Navy – The aircraft above was the U.S. Navy's second model of PB4Y and was designated the PB4Y-2. It was built by Consolidated Vultee for the U.S. Navy. Blending B-24 designs and the knowledge gained from the Navy's PB4Y-1 technology, the new aircraft was designed for low fuel consumption and more horsepower at takeoff. It had a top speed of 250 mph and a range that exceeded 3,000 miles. It was flown from land airbases. The aircraft's distinguishing features included a single taller tail assembly and side gun turrets housed in blisters off the fuselage which offered more protection. (Consolidated Vultee)

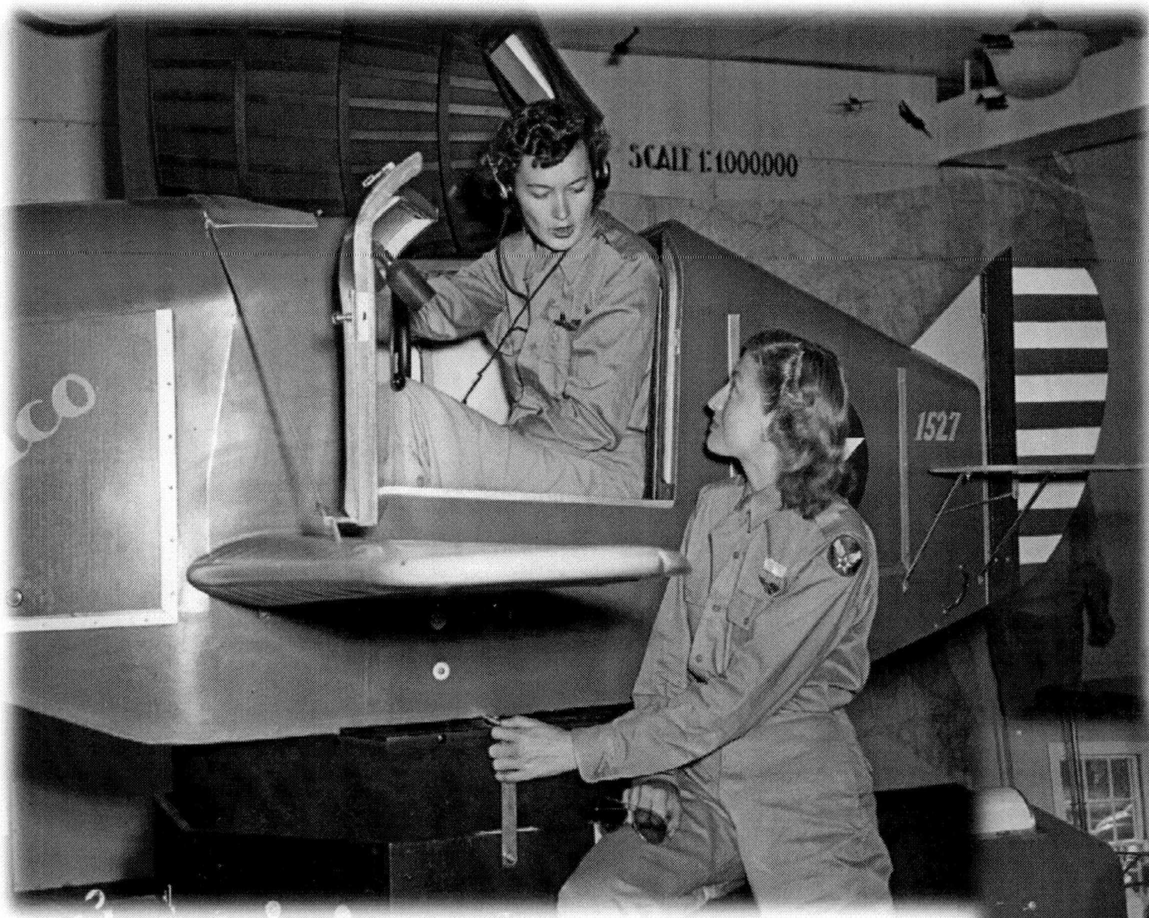

Link Trainer

Two women instructors, Caryl Jones in the cockpit and Mary Clifford beside the Link Trainer, were photographed as they discussed instrument flying. The Link Trainer was essentially a box with a hooded cockpit in the basic shape of an airplane. The procedure to learn to fly via the Link method had the student flyer seated in the cockpit, and the hood closed over their head. The instructor manipulated a series of instruments that made the simulator respond as if it were a plane in flight. By working certain levers and pushing various buttons, flight conditions were duplicated and the instrument panel in the Link revealed the status of the plane to the student pilot. (Jackie Martin)

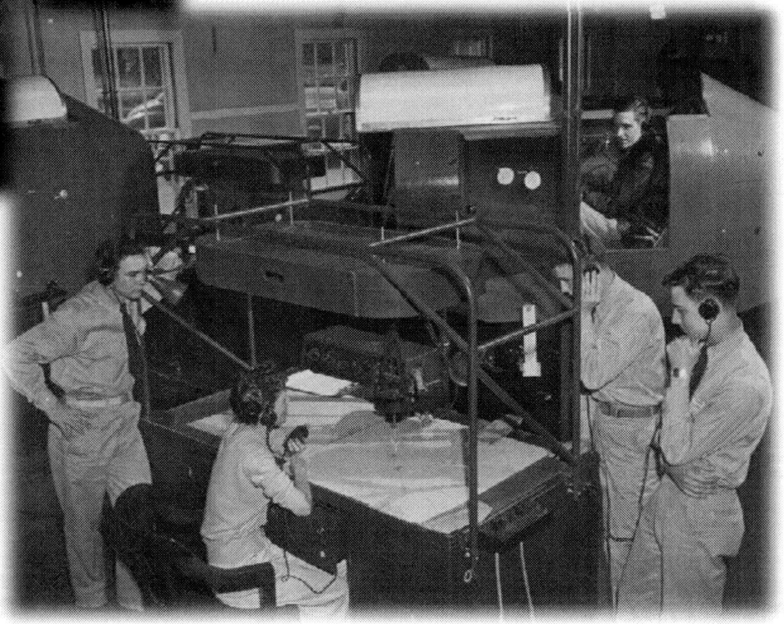

Link Trainer System – In the photograph to the right, Dorothy Wheeler instructed male pilots on how to fly solely on instruments in bad weather using the Link Trainer. The second largest customer for the Link Trainer was the Japanese Imperial Navy, who made a sizable purchase in 1935. Many Japanese pilots were trained in these devices and used their experience to find American targets. (U.S. Navy)

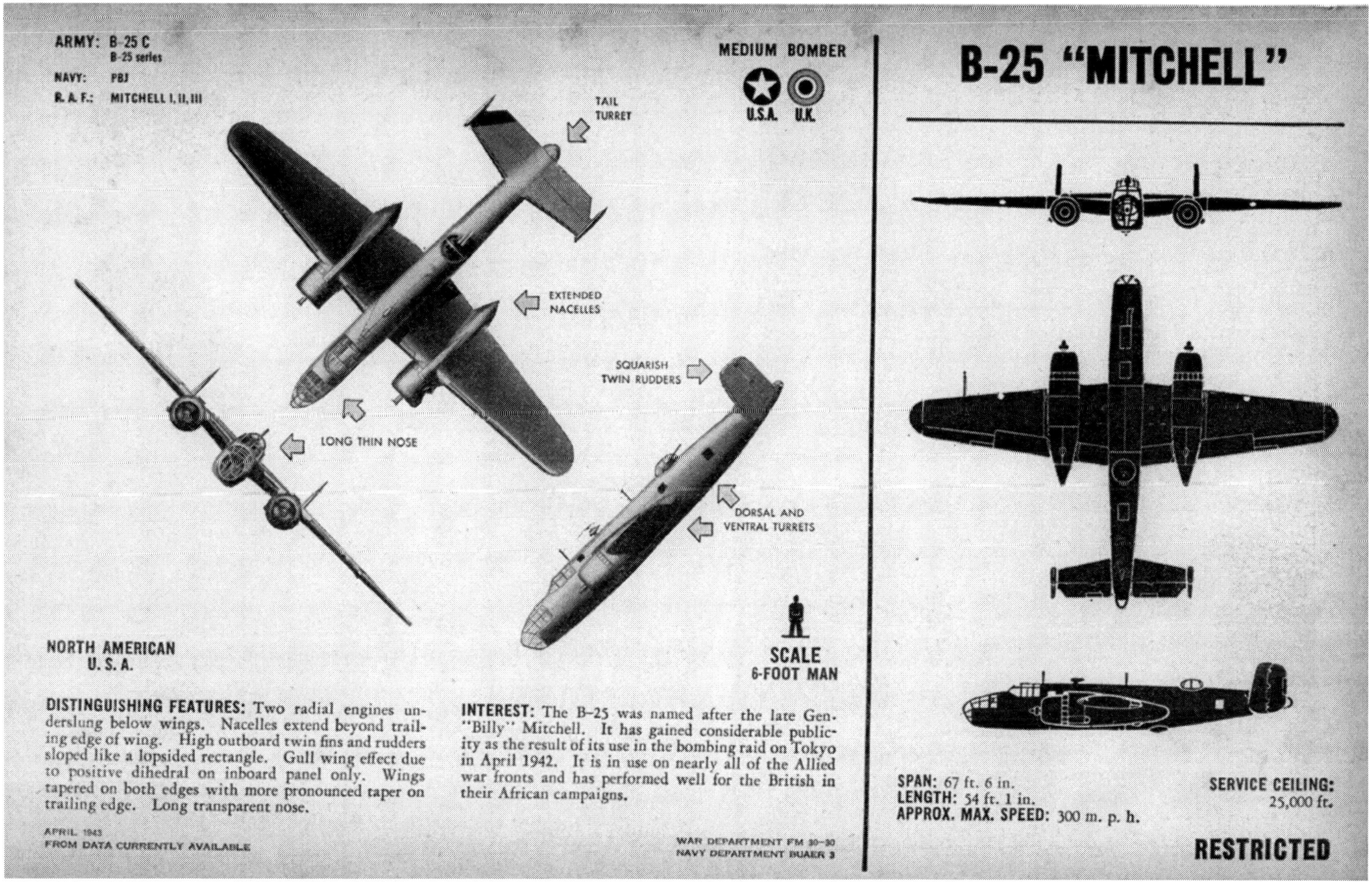

B-25 Mitchell – The North American B-25 Mitchell owes its beginnings to the Army's need for a medium bomber. Designed as a twin-tail, mid-wing monoplane, it was powered by two 1,700hp Wright Cyclone engines. Capable of more than 300 mph with a range of more than 3,000 miles when using drop tanks, it was used for high and low-level bombing, strafing, photo-reconnaissance, submarine patrol and even as a fighter.[1] It served as the aircraft that Jimmy Doolittle and 15 other pilots and crews flew from aircraft carriers to complete the historic surprise raid over Tokyo in 1942. Almost 10,000 were built from late 1939 through 1945. (War Department Recognition Guide 1943) [1]www.boeing.com/history/bna/b25.htm

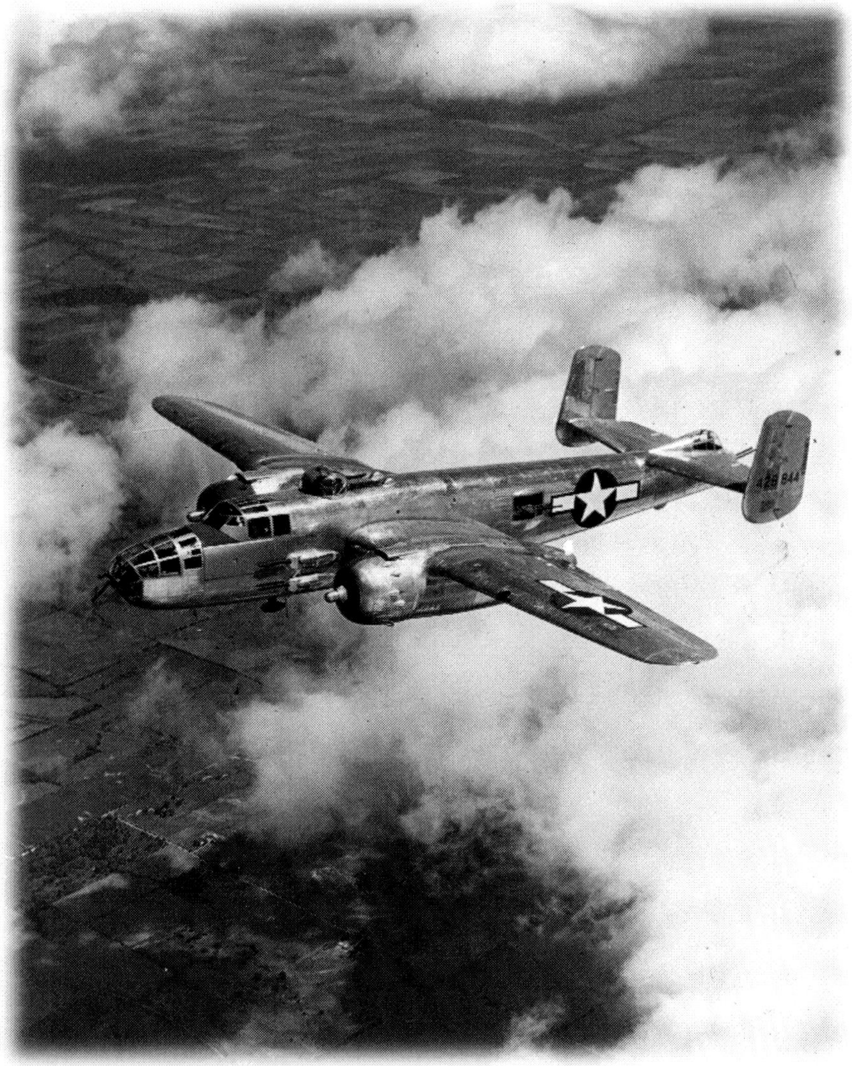

B-25J Mitchell – Famed for its superb ability to carry out low level attack and bombing missions, the B-25J Mitchell bomber, shown above, carried twelve .50 caliber machine guns for defensive and offensive purposes in addition to a load of bombs of about 4,000 pounds. (North American Aviation)

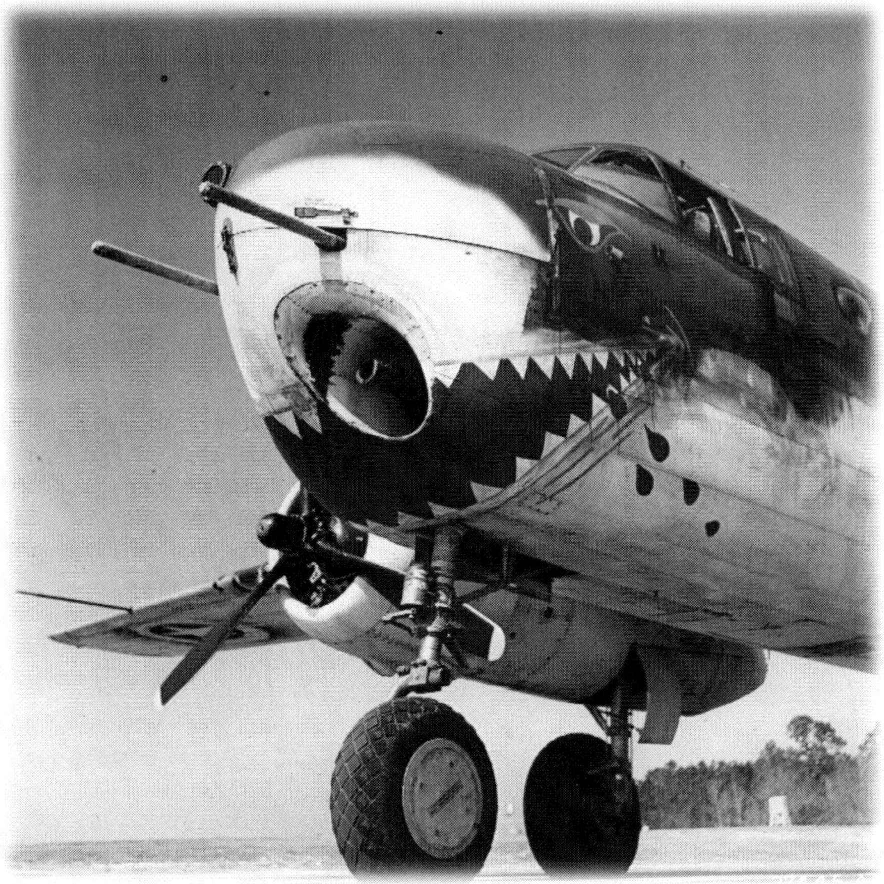

B-25 Mitchell – Looking at the business end of the upgraded B-25 shows a 75 mm gun set into the nose along with twin .50 caliber machine guns. At the time this photograph was taken, B-25s equipped with a 75 mm gun had already sunk many ships at sea. On the upper left of the nose, partially hidden by a machine gun muzzle, is a camera that could be used to film an attack. (U.S. Air Forces)

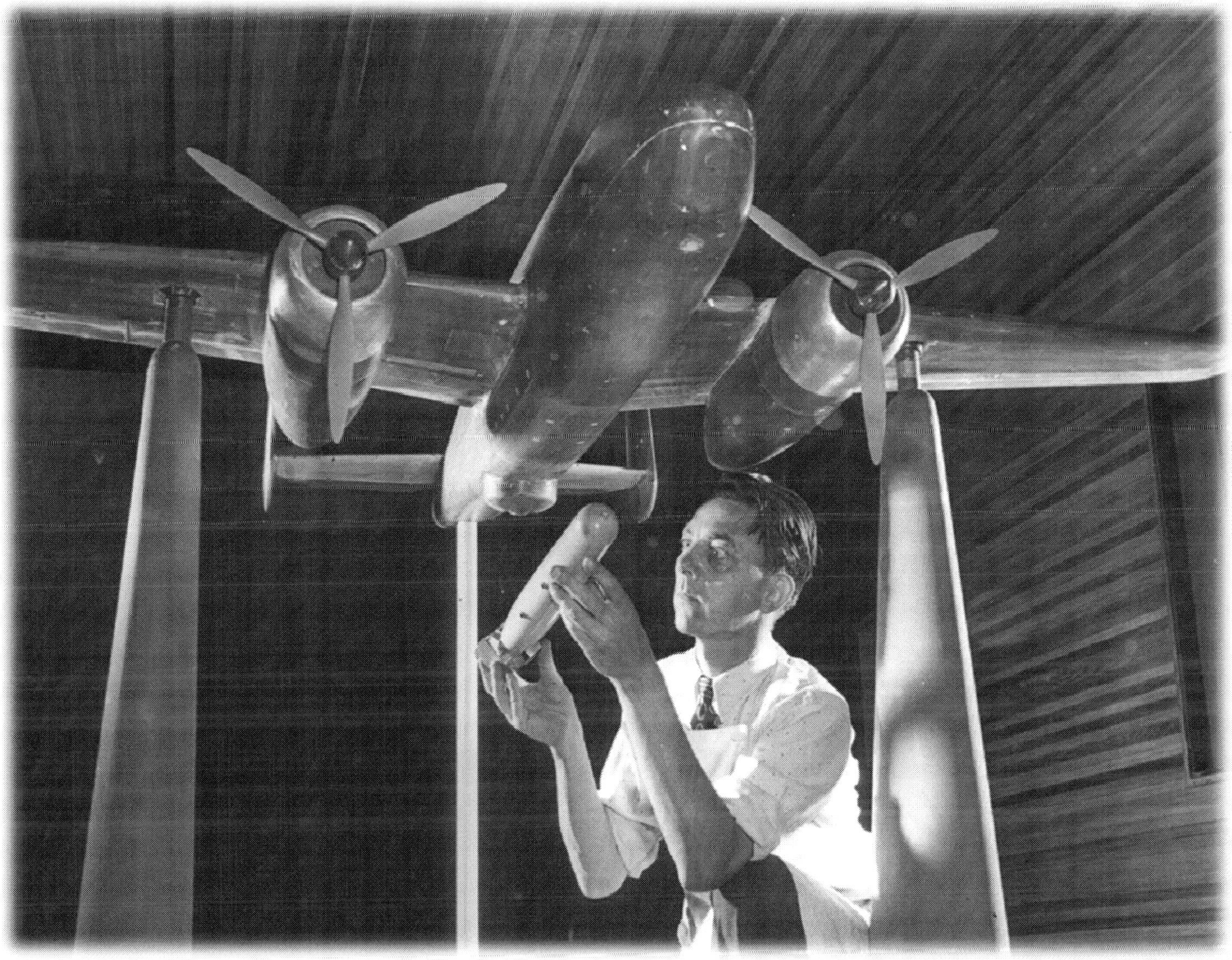

B-25 Mitchell – The B-25 was a versatile bomber and each new design extension required detailed tests. Computer simulations did not exist in WWII and engineers needed to conduct wind tunnel tests to understand the dynamics of the aircraft changes. Shown above, a North American designer prepared a scale model of a B-25 to test a method of carrying a torpedo. Eventually, a special rack for large torpedoes was adapted that was installed in the bomb bay. (North American Aviation)

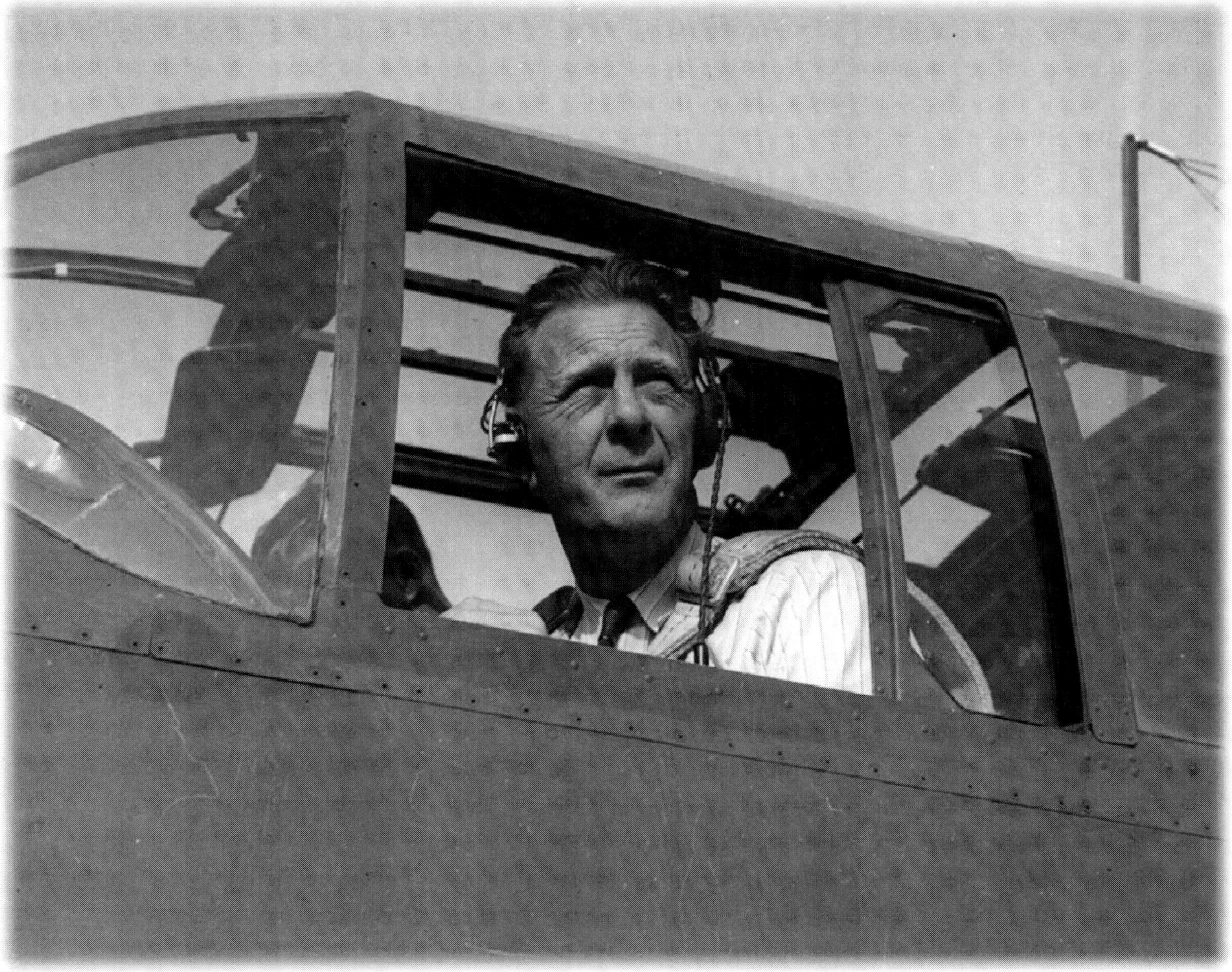

B-25 Mitchell – North American B-25 test pilot Roger Rudd was photographed in the pilot's seat of a B-25 that was equipped with a 75 mm gun. In WWI, Rudd was chief of a French 75 mm gun crew, the same size gun now used onboard this aircraft. Many different types of pilots flew B-25s including WASP (Women Airforce Service Pilots) that ferried the aircraft across the nation. (North American Aviation)

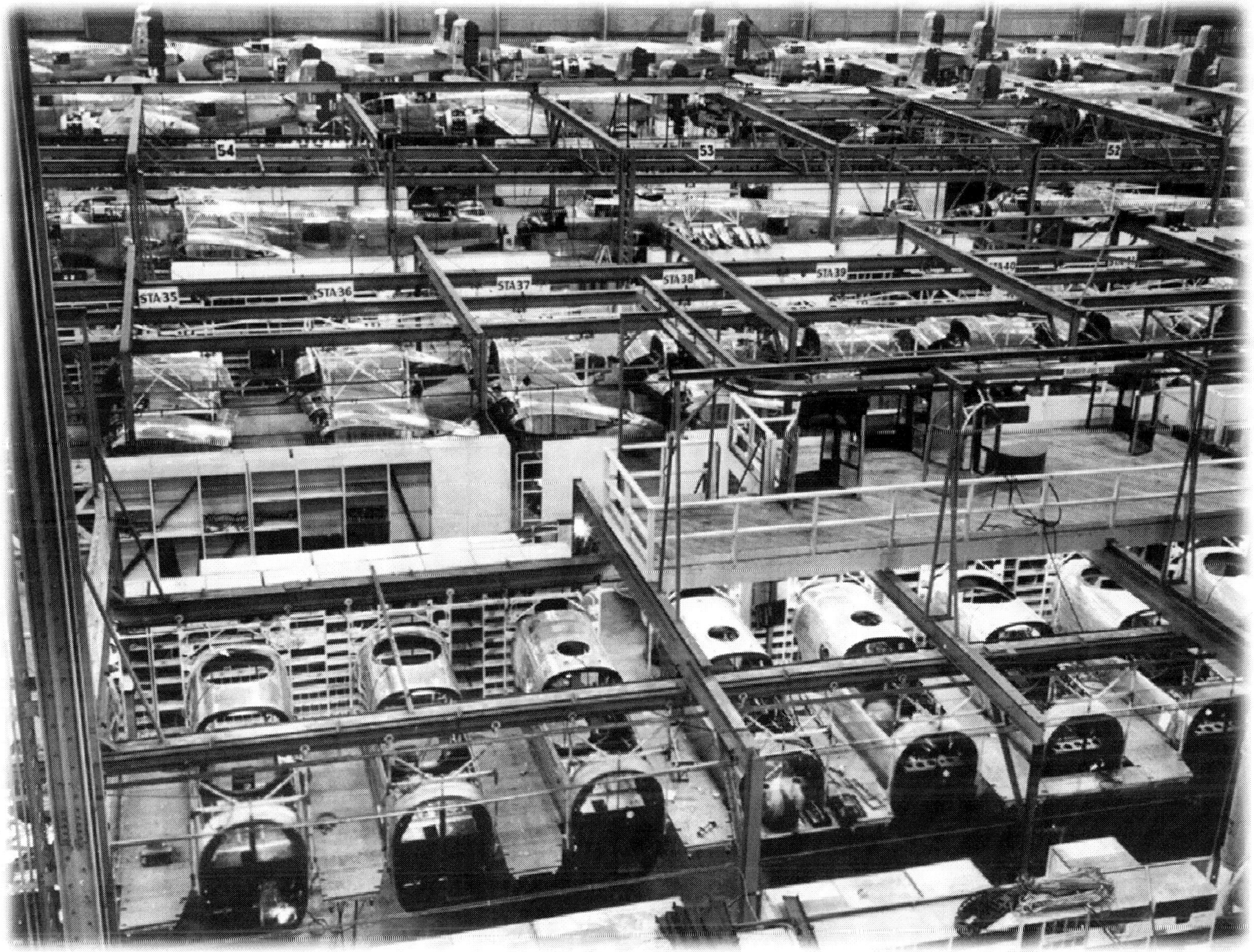

B-25 Mitchell – A bird's-eye view of the Kansas Division of North American Aviation shows the assembly conveyor lines building B-25s. Look closely and you will see the first line contains front sections while the second line is made up of center sections. The third line mated front, center and rear sections while the aircraft at the far back are nearly complete B-25s. (North American Aviation)

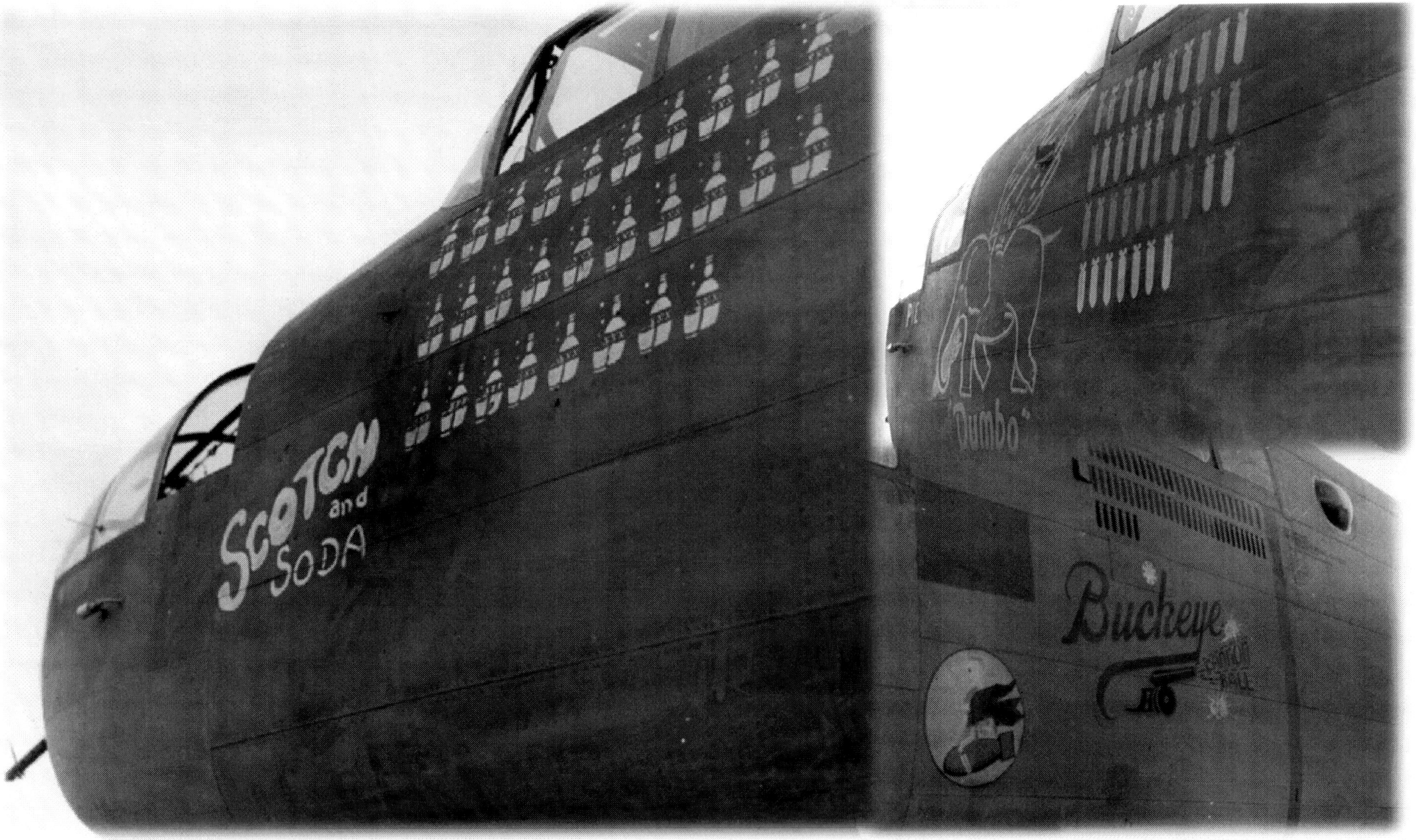

B-25 Mitchell – Take a look at how three pilots chose to customize their B-25s. Although these represent just three of the nearly 10,000 B-25s built, each portrays a unique crew personality. *Scotch and Soda* was piloted by Henry W. Stephenson from Atlanta, GA. At the time of the photograph, he had flown 28 missions and added another 31 by the time of the press release. *Dumbo* was piloted by Lieutenant Eugene E. Stocking from San Antonio, TX, who had 37 missions but added another 36 before the press release and last but not least, *Buckeye Cannon Ball* which was piloted by Captain Peter B. Soel from Athens, OH, with 56 missions in the photo but 83 total by the time of the press release. Each of these aircraft was part of the 12th Army Air Force serving in Italy. The press release is dated February 29, 1944. (U.S. Army Air Forces)

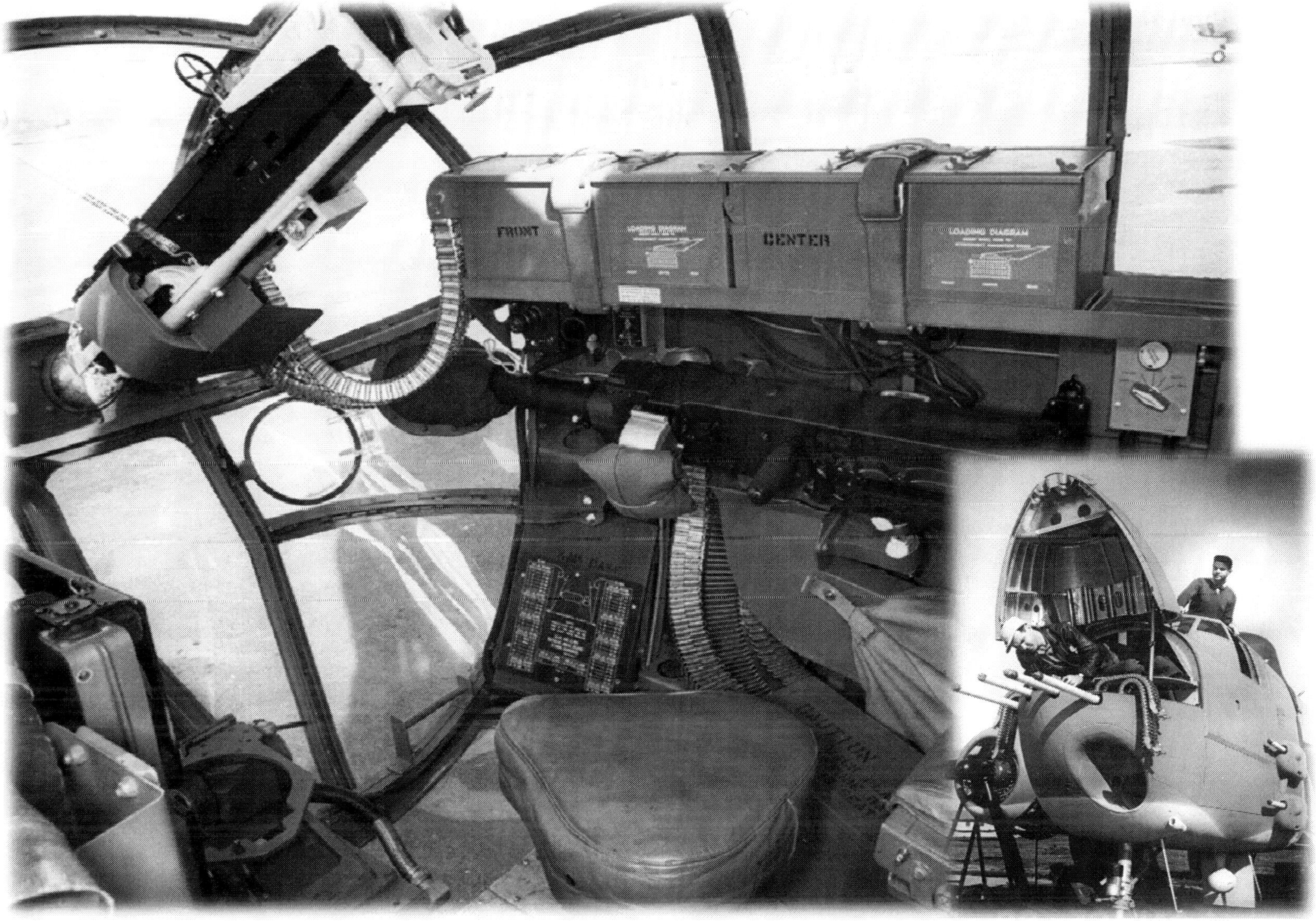

B-25 Mitchell – This North American Aviation photograph shows the flexible nose gun installation in the bombardier's compartment, with fixed gun. Both guns are .50 caliber. The nose design of a B-25 was altered many times depending on the model of the aircraft. This could be a B-25J which primarily functioned as a medium bomber and returned the nose to the transparent, bombardier-equipped nose of the earlier B-25C and D. The inset photo shows four .50 caliber guns and a 75 mm gun in a B-25 nose. (Both photographs by North American Aviation)

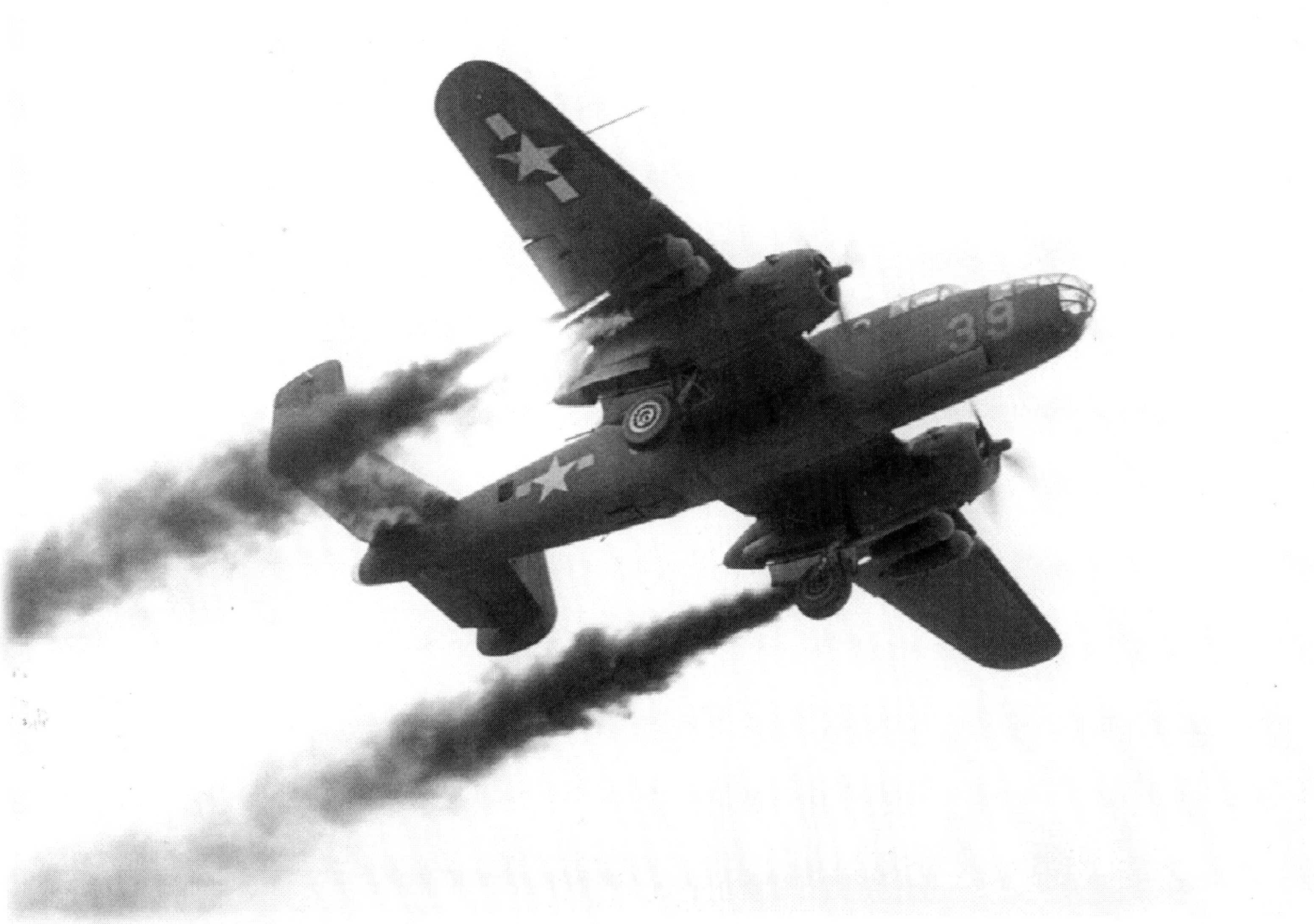

B-25 Mitchell – A rocket assisted B-25 was photographed testing shorter takeoffs and heavier bomb loads. The ram-jet engines were not put into tactical use on B-25s during WWII. The engines were being considered by the Army Air Force Material Command who had been testing rocket assisted takeoffs since 1940. This test was conducted at Wright Field, the headquarters of the command. (AAF Material Command)

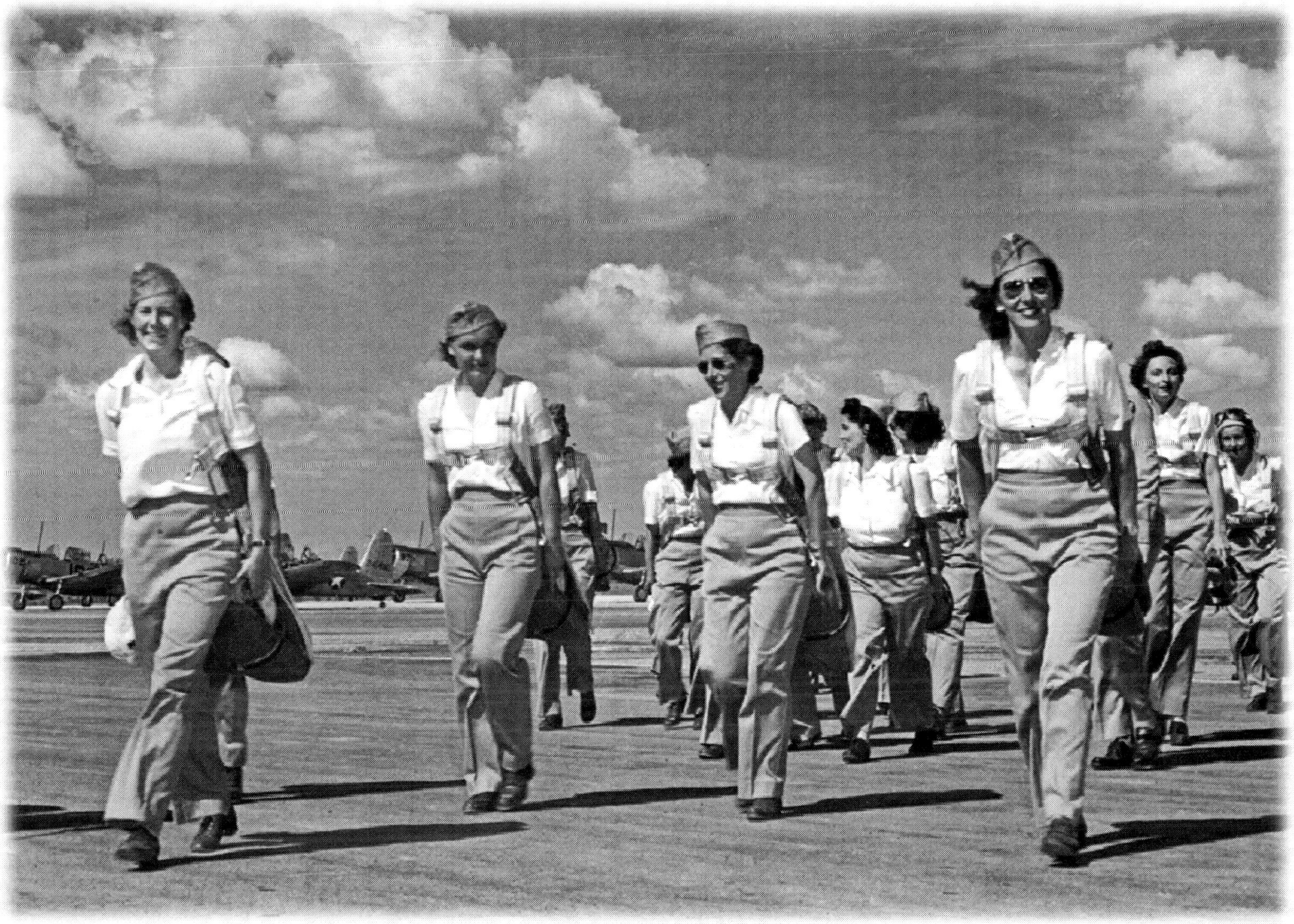

WASP Ferry Service – There was a serious shortage of pilots after December 7, 1941. Thousands of male American pilots were suddenly shipped overseas to fight. An equally important battle was brewing on the home front – how to move airplanes off the assembly lines and deliver them to military bases nationwide. There were simply not enough pilots. The shortage was compounded by the need to find replacement pilots who could face serious dangers including sabotage, rudimentary weather forecasting and flying new aircraft, barely tested, across long distances. To solve the problem, after some initial hedging, the U.S. Government launched an experimental program using women pilots eventually known as the Women Airforce Service Pilots or WASP. (Army Air Forces Training Command)

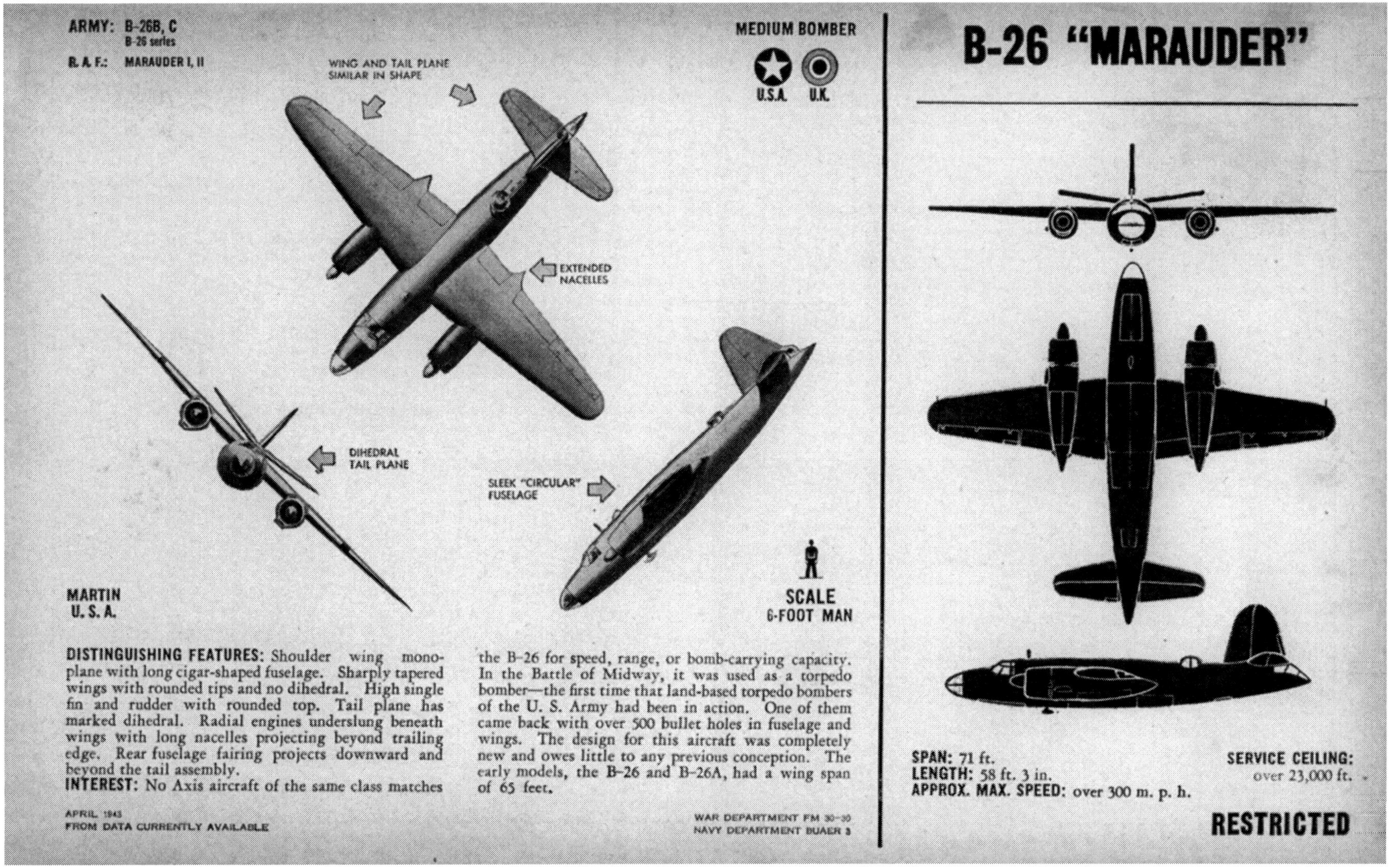

B-26 Marauder – The Martin B-26 Marauder flew missions in almost every theatre of operations during the war. The bomb load for the B-26 was between 2,000 – 4,000 pounds depending on target and range with a maximum theoretical bomb load of 4,800 pounds. It was rarely, if ever, carried in practice. Early versions were equipped with torpedo shackles on the underside of the fuselage to carry a single 2,000 pound Navy Mk.13 torpedo. The B-26 was equipped with two bomb bays, referred to as forward and aft bays. The rear bomb bay was never used for bomb carriage, and on later models this was sealed shut. Its twin 2,000hp Pratt & Whitney R-2800 Double Wasp radial engines allowed the aircraft to cruise at 216 mph. It flew from bases at Midway, the Aleutians, Guadalcanal, Australia, New Guinea, North Africa, the Mediterranean, France, England, Holland, and Belgium. Many Allied nations flew the bomber including the U.S. Army Air Force, U.S. Navy, British Royal Air Force, Free French Air Force and the South African Air Force. (War Department Recognition Guide 1943)

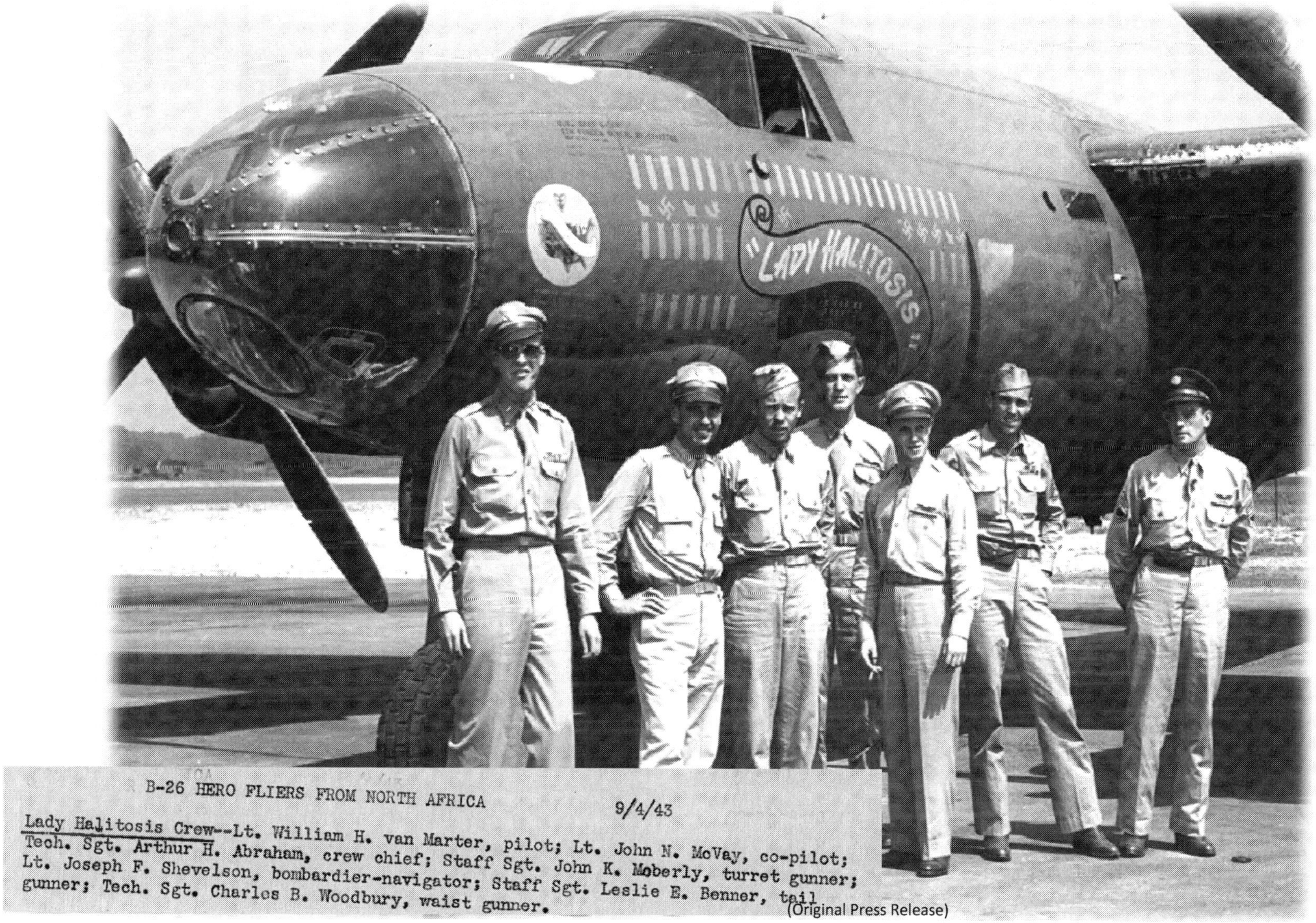

(Original Press Release)

B-26 Marauder – At the time of this photograph, the crew of *Lady Halitosis* had completed 43 bombing missions, destroyed six Nazi aircraft and sunk four ships. *Lady Halitosis* was serial 41-17765 and was one of three aircraft from the Mediterranean Theatre of Operations (one each from the 17th BG, 319th BG, and 320th BG) to be returned to the U.S. in July 1943 to participate in a "War Bonds Drive" to help raise funds for the war effort. The aircraft and her crew were photographed in the Fall of 1943. *Life Magazine,* July 10, 1944, featured the aircraft and crew among other heroes. Her pilot, Lieutenant William H. van Marter, survived the war and enjoyed a distinguished career in aviation. The aircraft was part of the 320th Bomb Group who flew from bases in Algeria, Tunisia, Sardinia and Corsica. (The Glenn L. Martin Company)

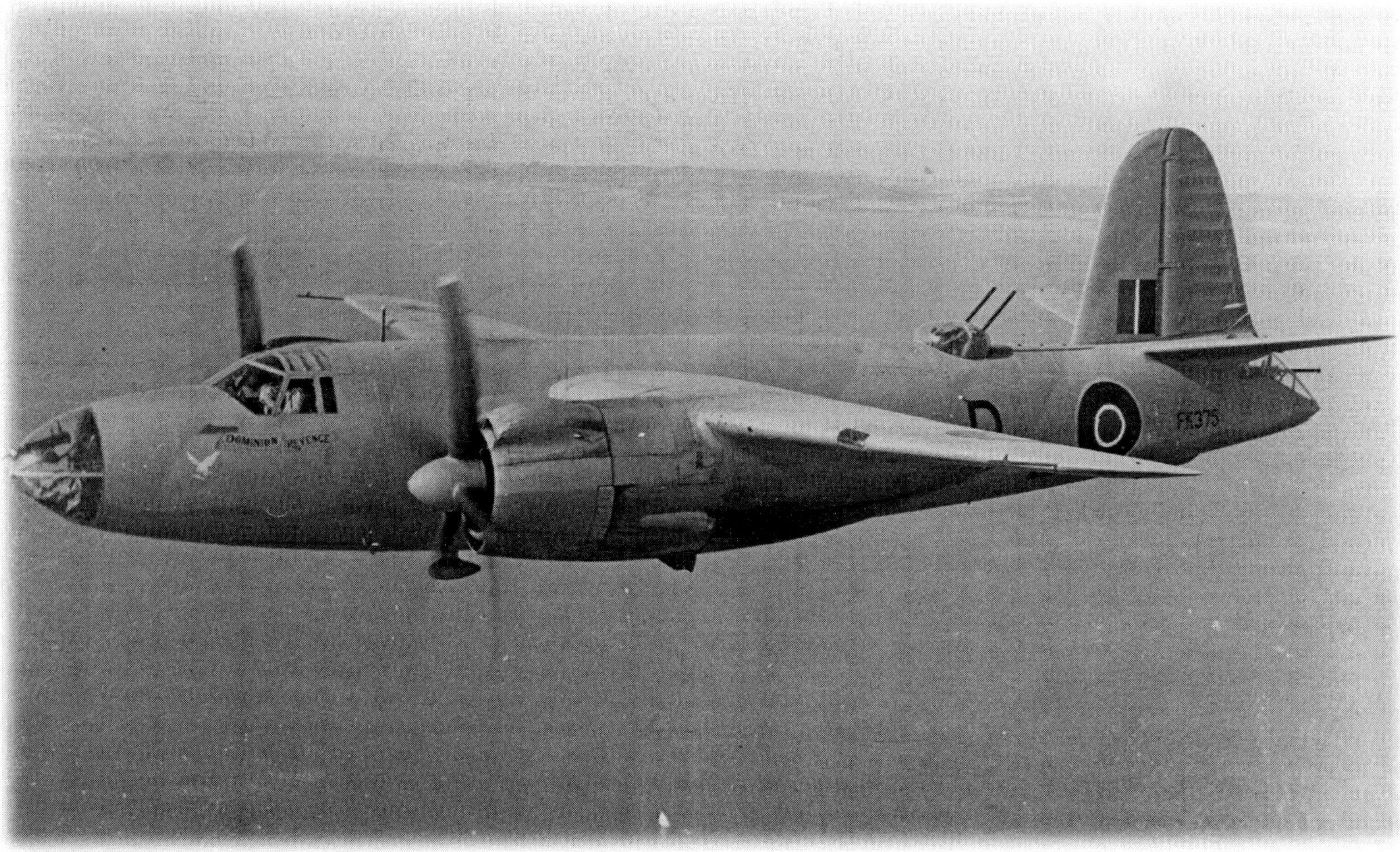

B-26 Marauder – British pilot Dick Maydwell was photographed flying an American built B-26 bomber supplied as a lend lease aircraft, and known in the RAF use as a Marauder Mk.I. He named it *Dominion Revenge*. He was the commander of the British 14 Squadron during 1942/43 stationed at Protville in Tunisia. He survived the war and when he was 90 years old he wrote, "We were flying at sea-level close to Cape Corse at the tip of Corsica, (in a different aircraft) when we saw the most enormous aircraft flying towards us. It was a six-engined enemy Me 323. I knew it had two formidable cannons firing aft so I maneuvered in front of the enemy aircraft and fired back at him. Soon three engines were out of action. The huge aircraft lumbered on towards the Corsican coast, where it crash-landed in a cloud of dust. Luckily for the crew, it didn't catch fire and no-one was injured as the gun crews and tractor drivers had all gone to the rear of the airplane. We did not shoot them up on the ground. I have been friends with the pilot of the Me 323 for the last 21 years, but that is another story!"[1]

(British Information Services) [1]www.14sqn-association.org.uk/14_Squadron_Association/Memoirs.html

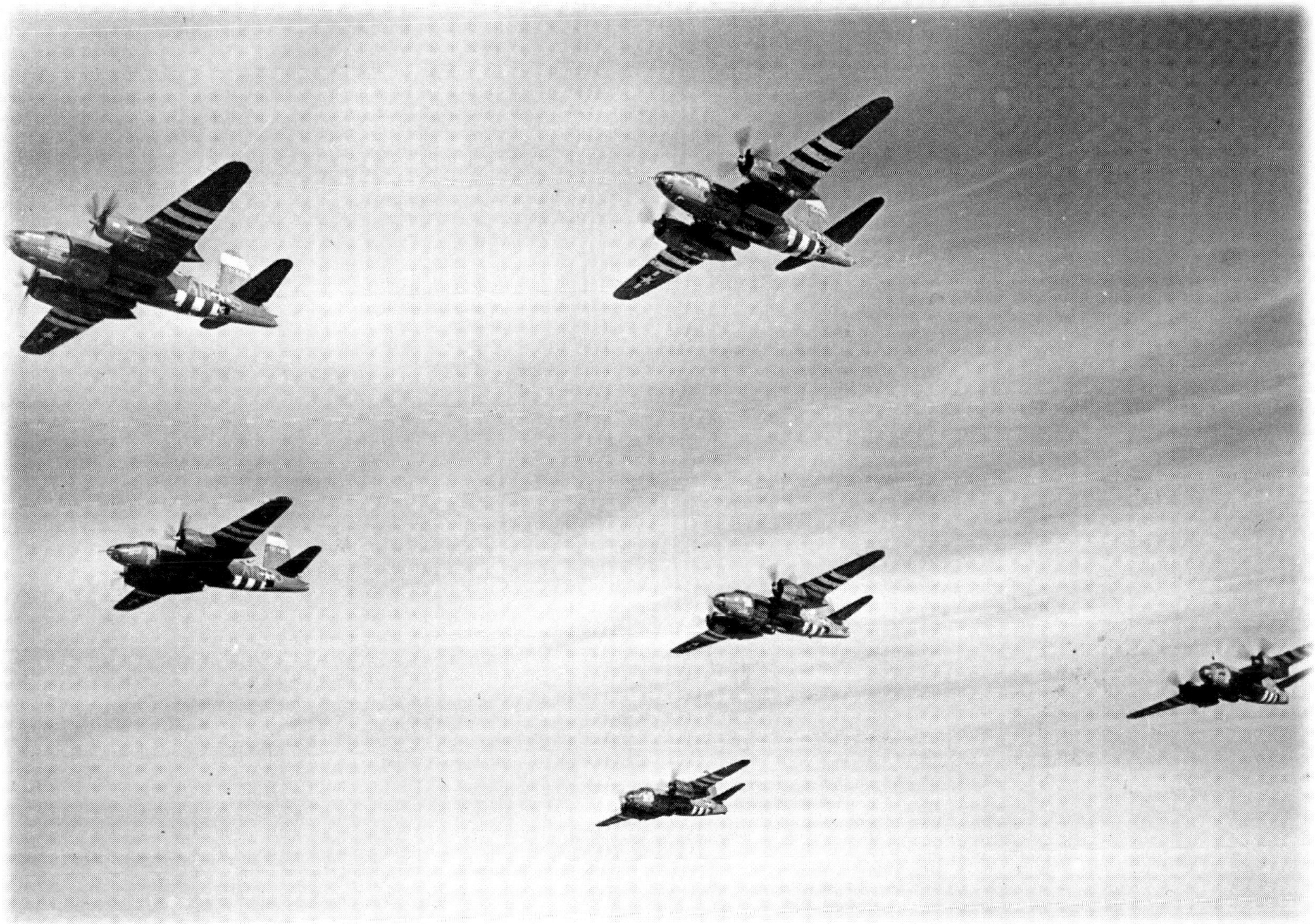

B-26 Marauder – This photograph captured some of the flight of 600 B-26s that flew in D-Day, the invasion of France. Noted on the back of the photograph is that each of these Marauders had completed at least 100 missions before June 6, 1944. All of the aircraft flew from the 386th Bomb Group of the 9th Air Force commanded by Colonel Joe W. Kelly. The aircraft pictured are *Litljo, Rat Poison, Bar Fly, Ye Olde Crock, Blazing Heat* and *Slightly Dangerous*. (U.S. Army Air Forces)

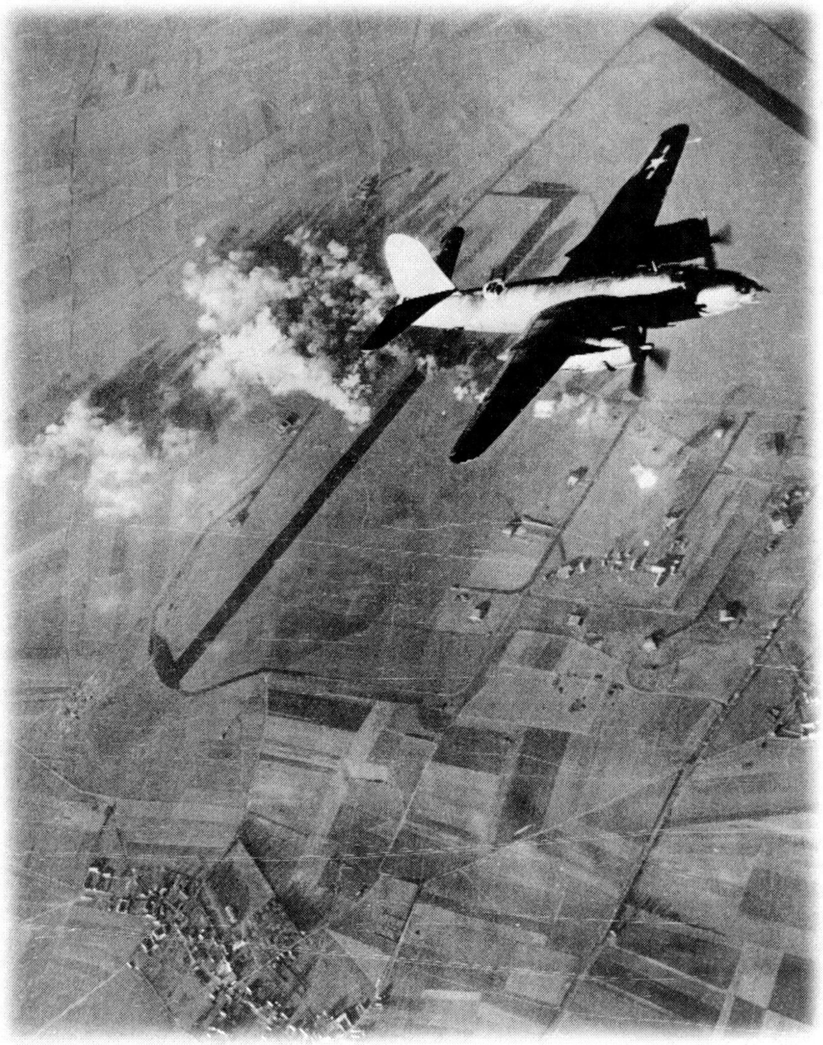

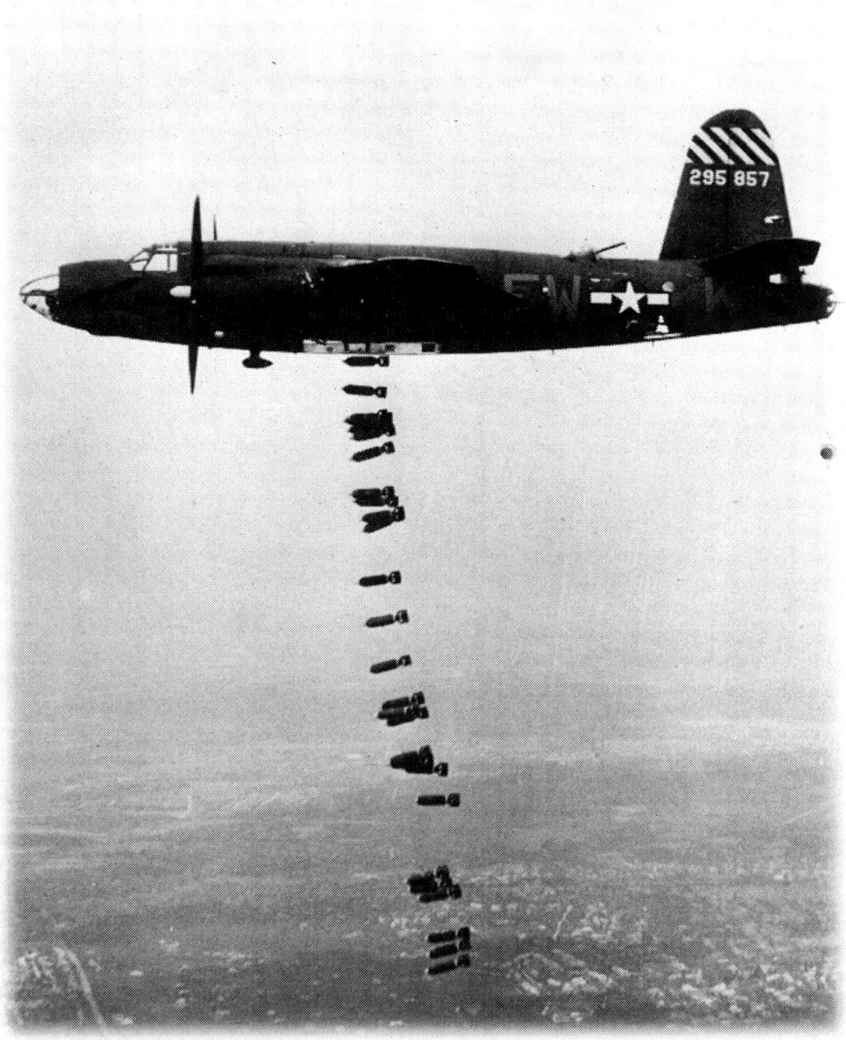

B-26 Marauder – Above left, B-26 bombers from the 8th Air Force attacked a German bomber base, Cambrai-Niergries, in France on December 1, 1943. On the right, *Shootin' In,* a B-26 bomber from the 9th Air Force, was photographed releasing more than twenty-six 100 pound bombs on another target in France. *"Shootin' In"* served with the 387th Bomb Group / 556th Bomb Squadron. It was named *Shootin' In* by pilot Lt. Asa V. Clark, after a phrase the crew often used when taking an overnight pass to London, "Shoot'in in to London". The B-26 medium bomber had a 1,150 mile range carrying a bomb load. Fast and effective, the aircraft flew more than 110,000 sorties and dropped more than 150,000 tons of bombs.[1] Over 5,260 B-26s were produced before production was stopped in 1945. (Both photographs U.S. Army Air Forces)

[1]"Martin B-26G Marauder." airforcehistory.hq.af.mil.

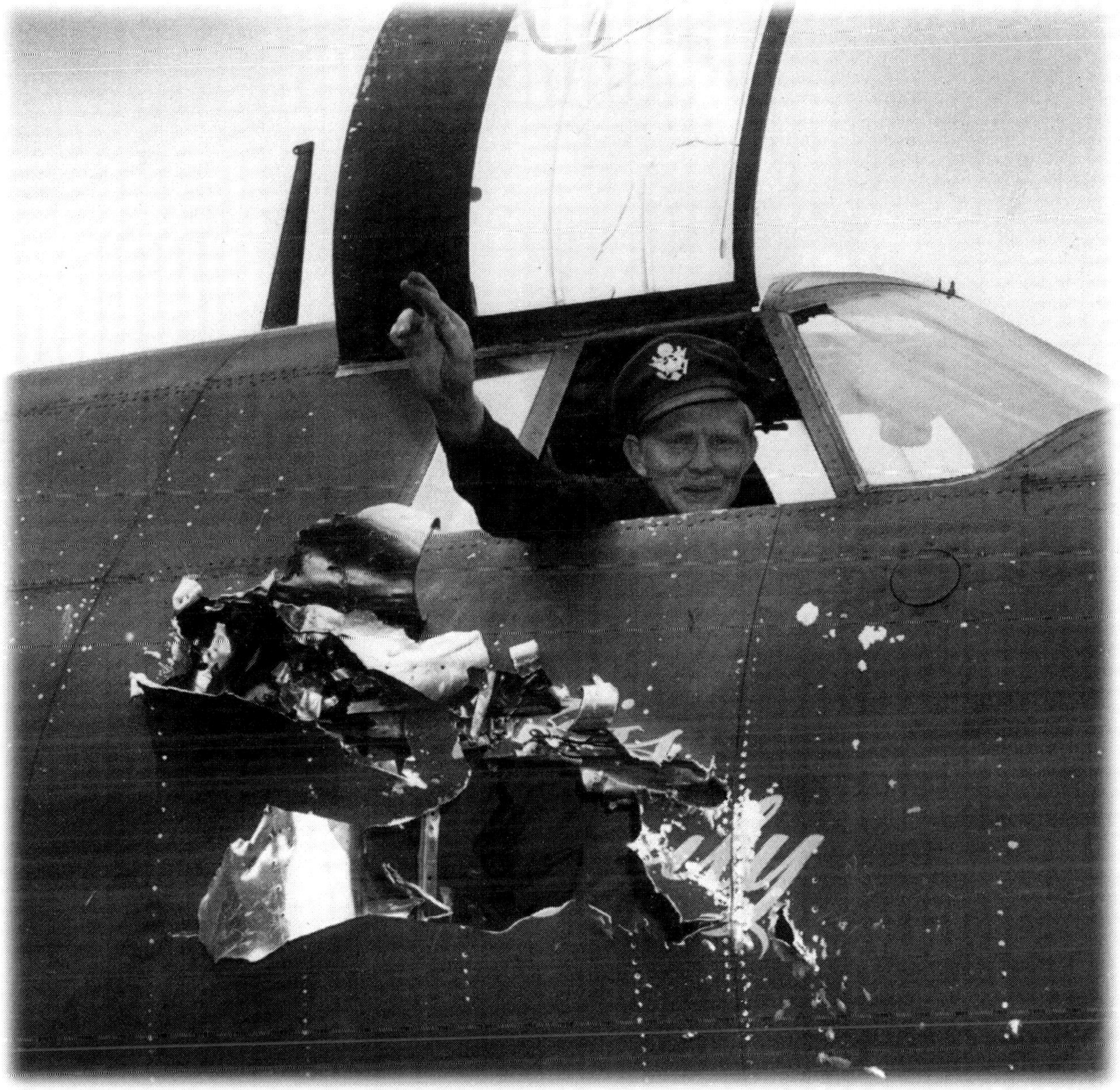

B-26 Marauder

The handwritten note on the back of this photograph states, "This Marauder took it from Nazi 20 mm cannon, but returned safely to its base in England."

The damaged aircraft was *Miss Emily*, serial 41-34719, serving with the 323rd Bomb Group / 455th Bomb Squadron. It was hit by flak on August 2, 1943 during a mission to Merville Airfield, France. A flak shell passed through the lower fuselage and blasted out through the cockpit (as seen). Luckily no one was seriously wounded, including the pilot, Lt. James Davis.

This variant of the Marauder, used exclusively in combat by the 323rd Bomb Group, was designated as a B-26C-6-MO, and was flown by a single pilot (not the usual pilot and co-pilot as on all other B-26 variants).

A pilot from the 344th Bomb Group said, "The Martin B-26 wasn't a bad plane when I flew it, so I would have to say a big NO to calling it Flying Prostitute or Widow Maker. Another old name that I have heard was Baltimore Whore."[1]

(The Glenn L. Martin Company)

[1]www.b26.com/page/shop_talk.htm

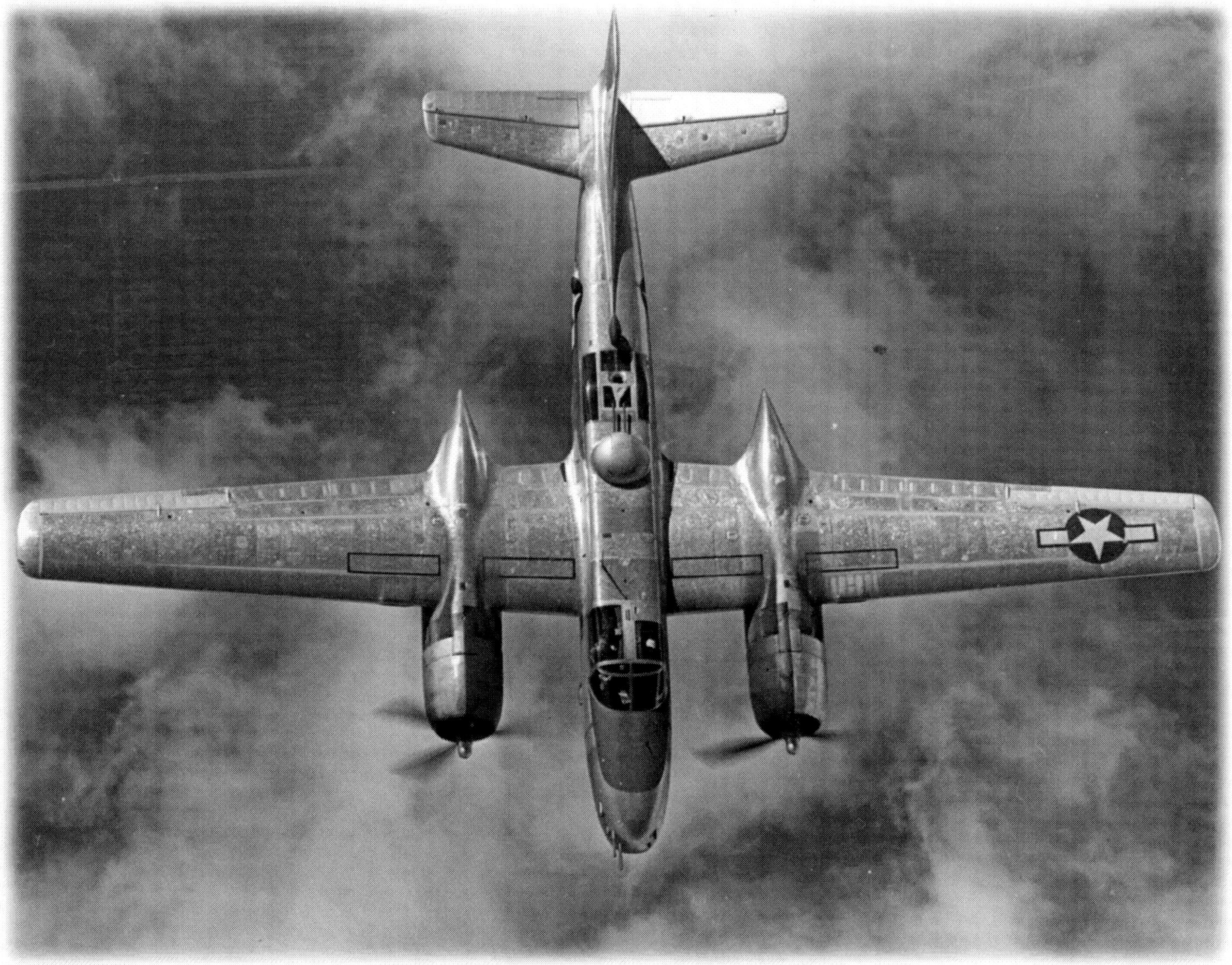

A-26 became the B-26 Invader – There was a second B-26 bomber. After the B-26 Marauder was retired from service in June 1948, the Douglas A-26 Invader was re-designated B-26. It was considered a light bomber, hosted a 70 foot wing span and carried 6,000 pounds of bombs. The A-26 was successful in WWII where, in the 9th Air Force, it flew 11,567 missions, dropping 18,054 tons of bombs.[1] The aircraft and crews fought tough battles in Korea with initial targets south of the 38th parallel selected to prevent reinforcement of the enemy forces. (International News Photo) [1]Mesko, Jim. *A-26 Invader in Action* (Aircraft Number 37) Carrollton, Texas: Squadron/Signal Publications, 1980. ISBN 0-89747-093-1

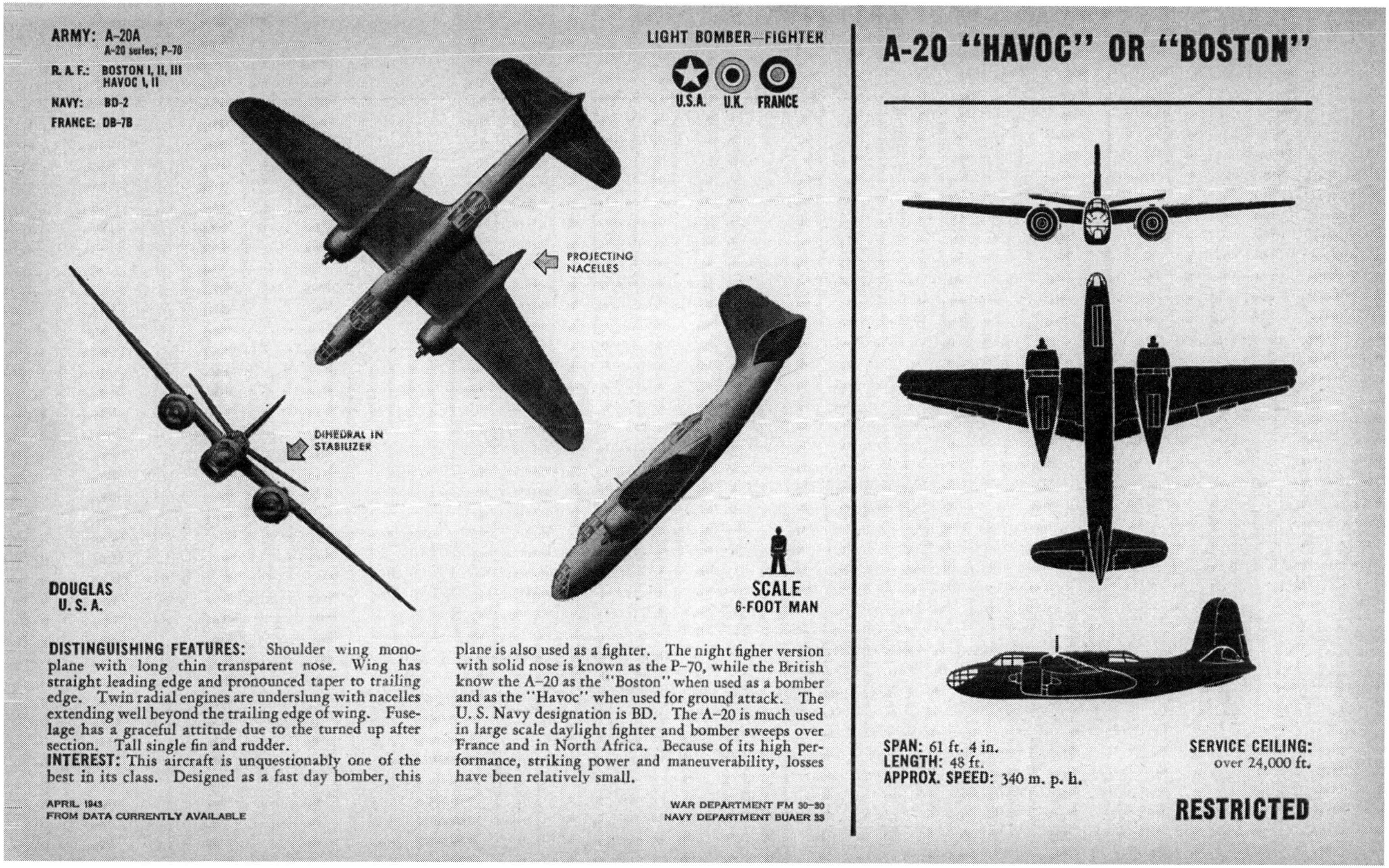

ARMY: A-20A
A-20 series; P-70
R. A. F.: BOSTON I, II, III
HAVOC I, II
NAVY: BD-2
FRANCE: DB-7B

LIGHT BOMBER—FIGHTER

U.S.A. U.K. FRANCE

A-20 "HAVOC" OR "BOSTON"

PROJECTING NACELLES

DIHEDRAL IN STABILIZER

DOUGLAS
U. S. A.

SCALE
6-FOOT MAN

DISTINGUISHING FEATURES: Shoulder wing monoplane with long thin transparent nose. Wing has straight leading edge and pronounced taper to trailing edge. Twin radial engines are underslung with nacelles extending well beyond the trailing edge of wing. Fuselage has a graceful attitude due to the turned up after section. Tall single fin and rudder.
INTEREST: This aircraft is unquestionably one of the best in its class. Designed as a fast day bomber, this plane is also used as a fighter. The night figher version with solid nose is known as the P-70, while the British know the A-20 as the "Boston" when used as a bomber and as the "Havoc" when used for ground attack. The U. S. Navy designation is BD. The A-20 is much used in large scale daylight fighter and bomber sweeps over France and in North Africa. Because of its high performance, striking power and maneuverability, losses have been relatively small.

SPAN: 61 ft. 4 in.
LENGTH: 48 ft.
APPROX. SPEED: 340 m. p. h.
SERVICE CEILING: over 24,000 ft.

APRIL 1943
FROM DATA CURRENTLY AVAILABLE

WAR DEPARTMENT FM 30-30
NAVY DEPARTMENT BUAER 33

RESTRICTED

A-20 Havoc – The A-20 was a WWII light bomber often operating as a heavy fighter. Produced by Douglas Aircraft Company, the aircraft had a wingspan of 61 feet 4 inches and was almost 48 feet long. With a maximum speed of 339 mph and a range of just over 1,000 miles, the light bomber could deliver 4,000 pounds of ordnance quickly. The aircraft was replaced during WWII by the Douglas A-26 Invader, the P-70 and the Northrop P-61 Black Widow serving as a night-fighter. Many flyers of the aircraft think it was one of the most versatile bombers of WWII. (War Department Recognition Guide 1943)

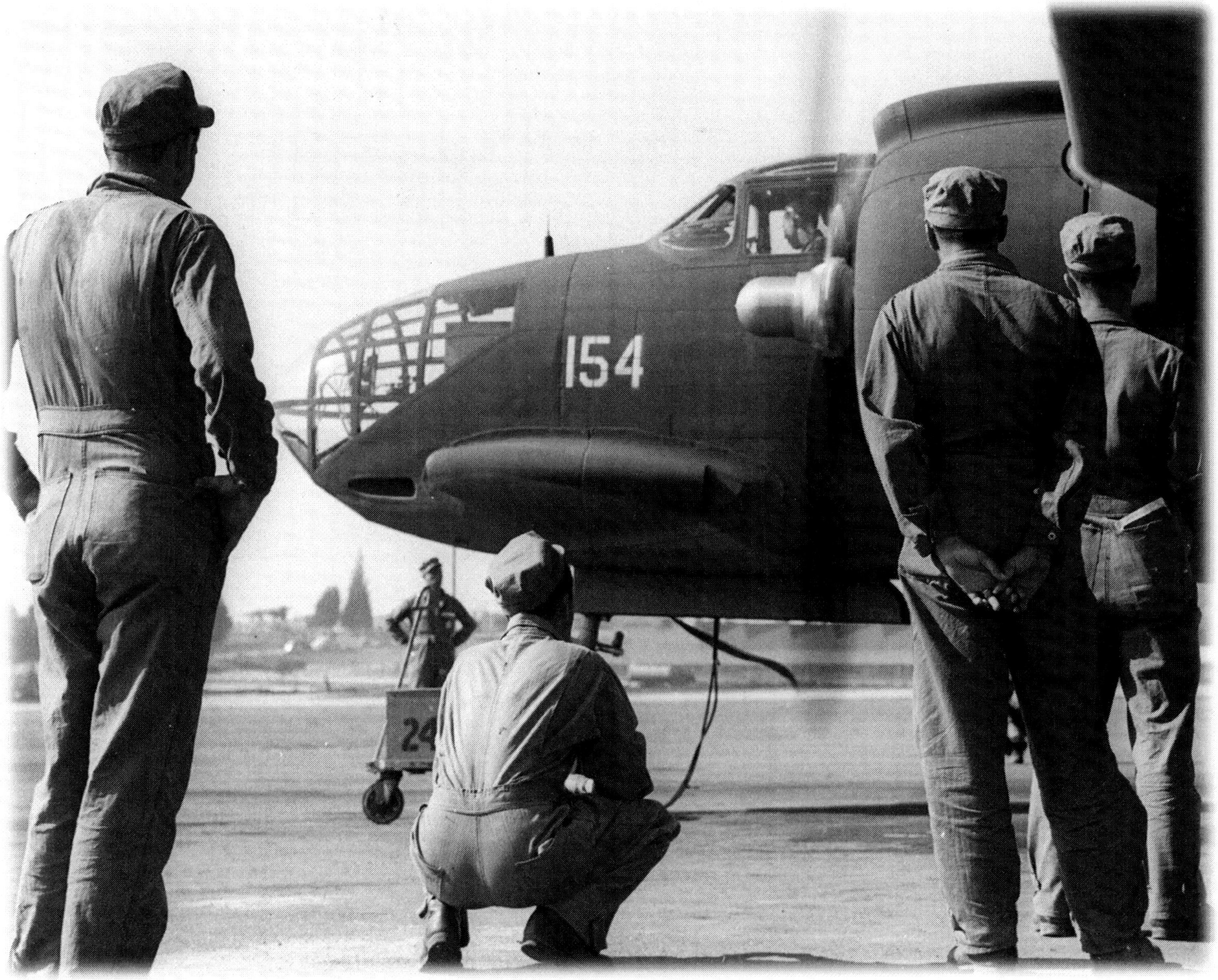

A-20 Havoc – The ground crew of this A-20 listened to the new engines that had just been installed on the aircraft. Early models of the A-20 saw several improvements in power and engine designs. (U.S. Army Air Forces)

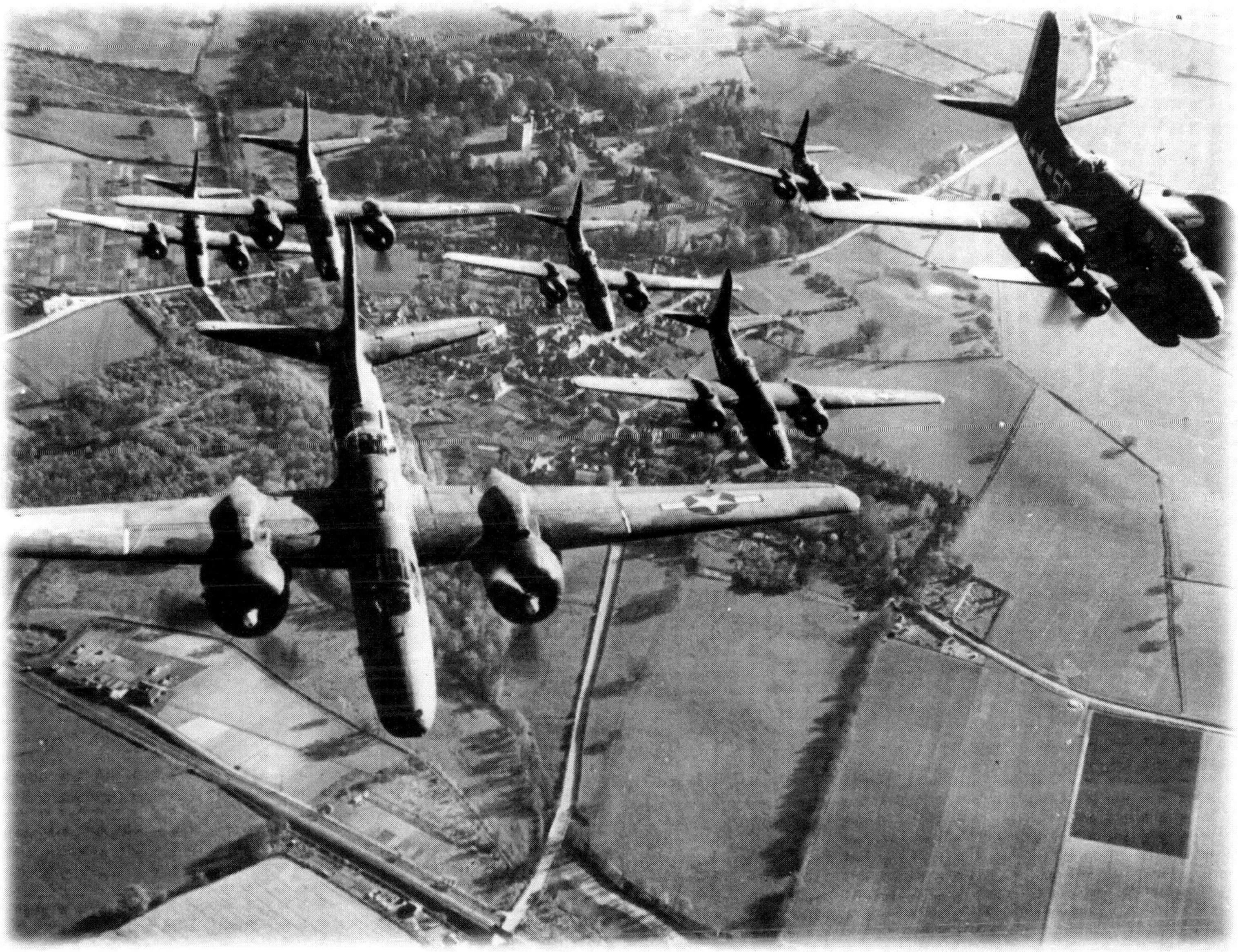

A-20 Havoc – The aircraft photographed above were based in the European Theatre and had just finished bombing an ammunition storage facility near the coastal town of Dieppe, France. It was noted, no aircraft were lost on the mission. A total of 7,478 A-20s were built in various designations and more than half were flown by other nations. The A-20 was a loud aircraft and troops on the ground often sited that they could hear the bomber from more than a mile away. About 40% of A-20 crews were assigned to the Pacific Theatre. (U.S. Army Air Forces)

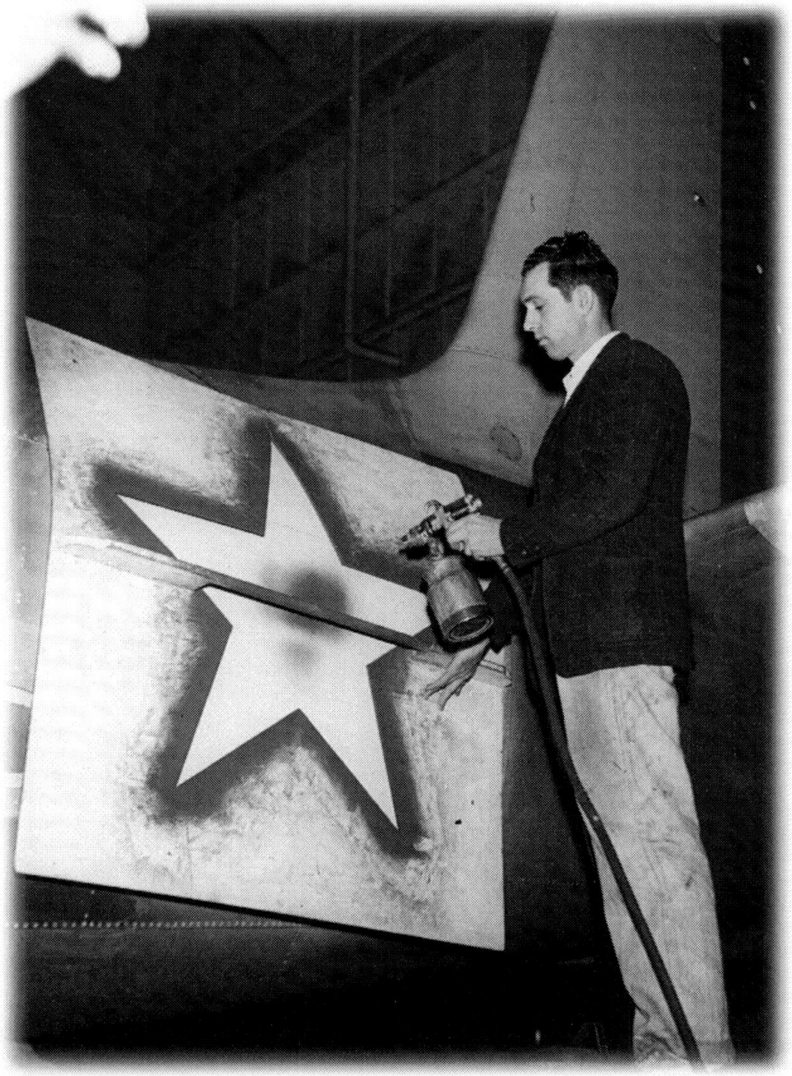

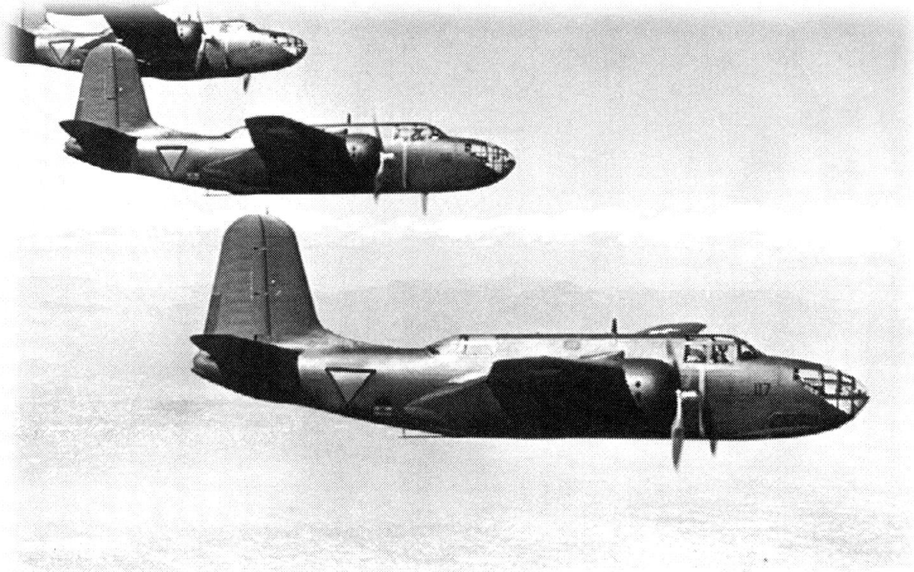

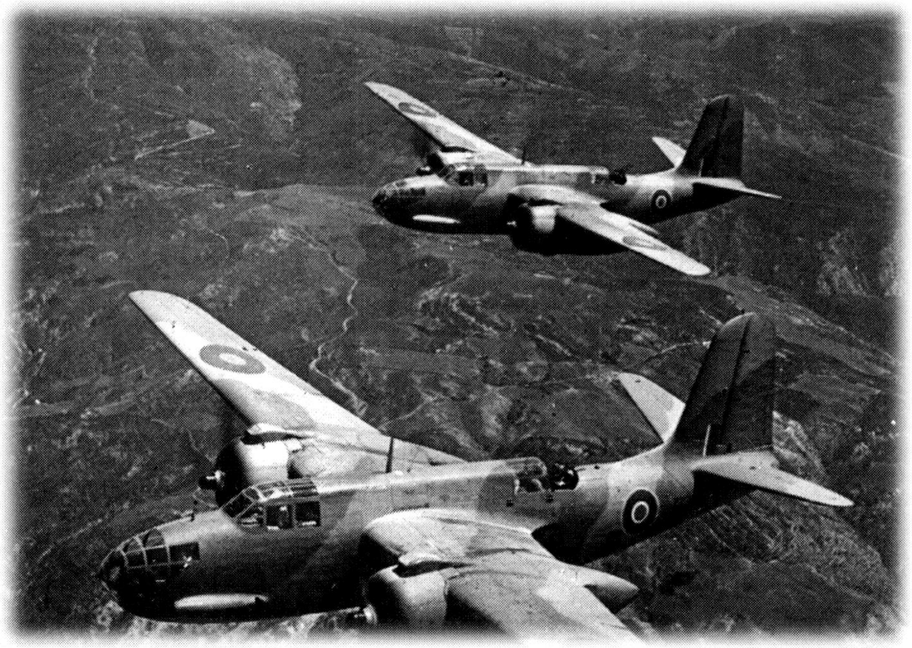

A-20 Havoc – The Lend-Lease program provided A-20 bombers to many nations. Top left, the Red Star of Russia was being painted onto an A-20 prior to shipment overseas. Photographs on the right show A-20s in both Dutch (top) and British (bottom) colors. Lend-Lease was an Act of Congress that provided that the President could ship supplies, including weapons, to any country fighting the Axis enemy that would assist in U.S. defense. (Douglas Aircraft Company/Netherlands Information Bureau/British Information Services)

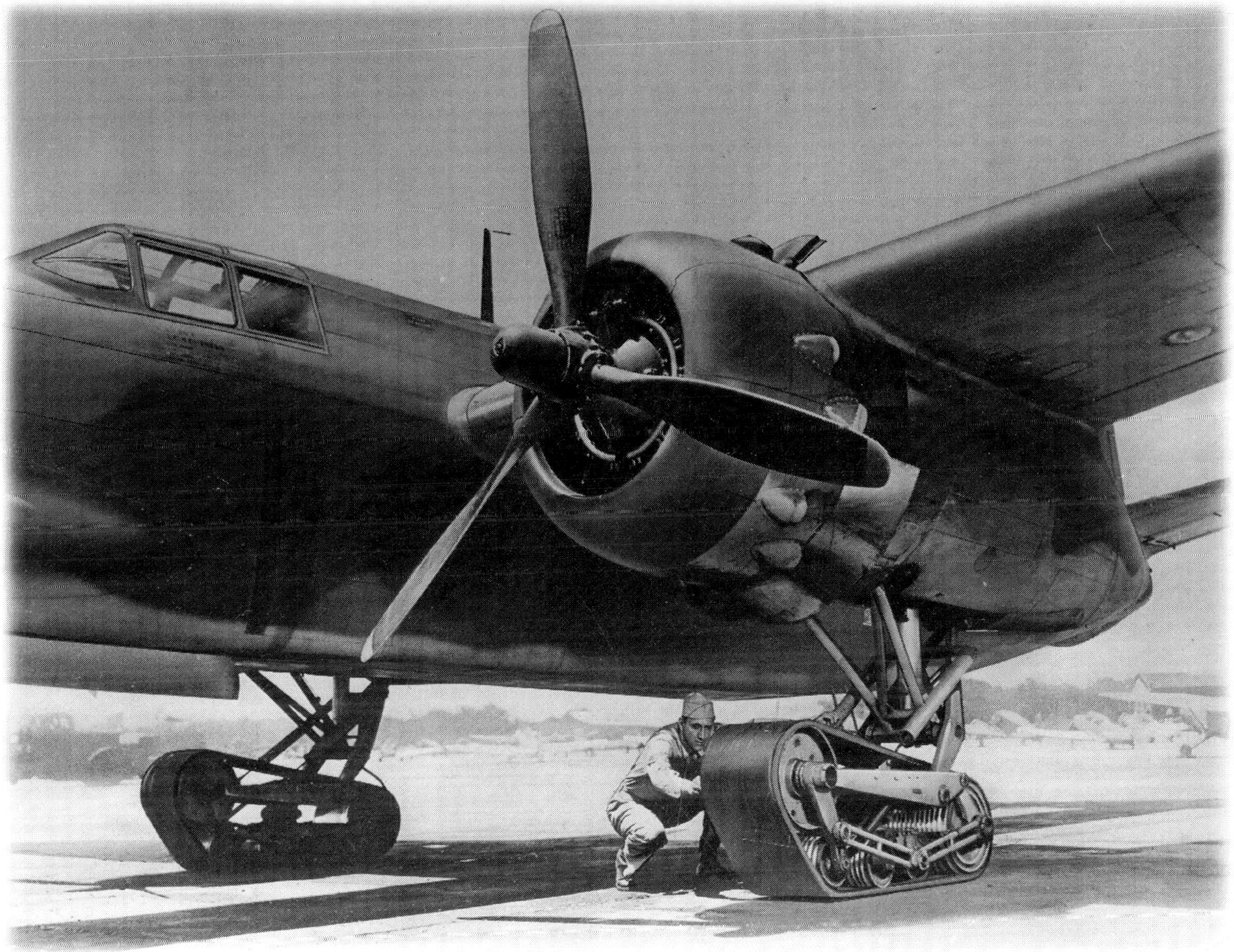

A-20 Havoc – Caterpillar-type landing gear was designed experimentally for the A-20 Havoc and tested at Wright Field, a U.S. Army aircraft laboratory. Designers found that the new gear permitted the aircraft to land on soft, sandy or excessively rough ground at speeds as high as 120 mph. Important in its day, the tread used about the same amount of rubber as conventional gear. It's unclear if any production bombers were fitted with the landing gear. (U.S. Army Air Forces)

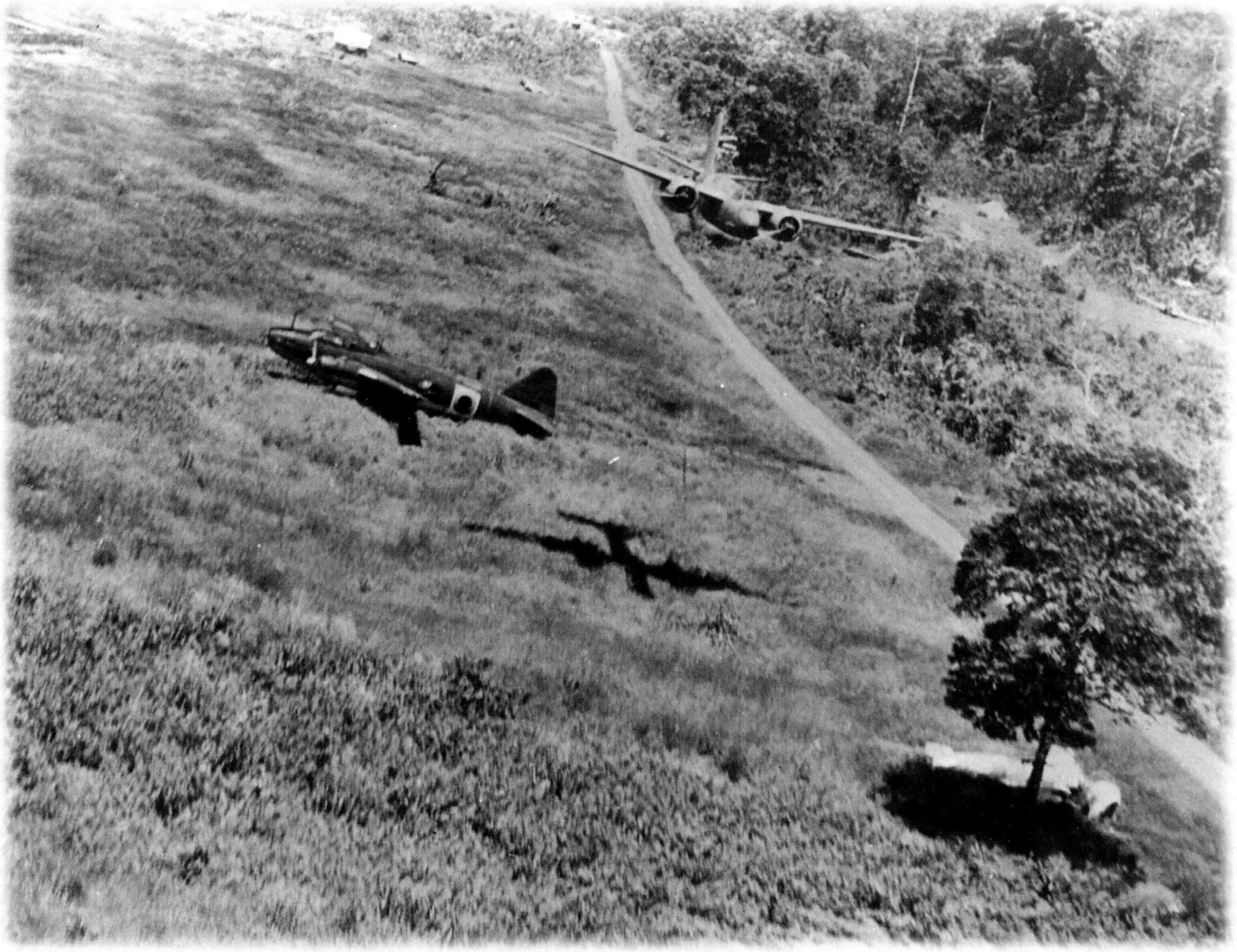

A-20 Havoc – Flying 100 feet over New Guinea, this A-20 was in the process of strafing Japanese planes on the ground. Noted in pencil on the reverse of the photograph, a spotter highlighted that five aircraft were attacked; two near the house at the top of the photo, a twin engine aircraft in the center field, a Zero under a tree, bottom right, and one off to the right in a small clearing. (Douglas Aircraft Company)

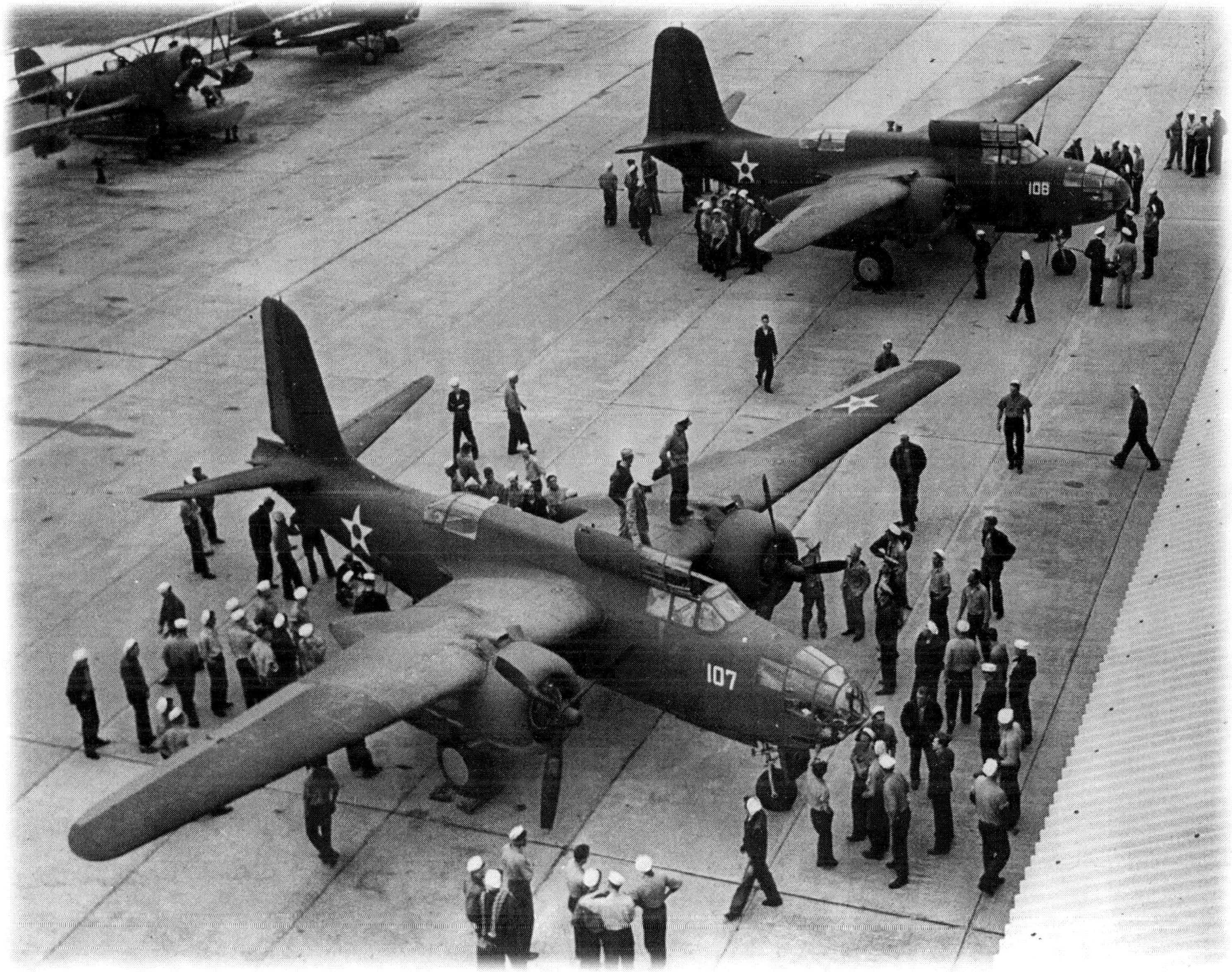

A-20 Havoc – Look closely at the servicemen standing around and on top of these A-20s and you will see sailors. These aircraft were going to the U.S. Navy to be flown as land based bombers to protect shore bases. The aircraft was effective at protecting U.S. interests from any enemy carrier based fighters. (U.S. Navy)

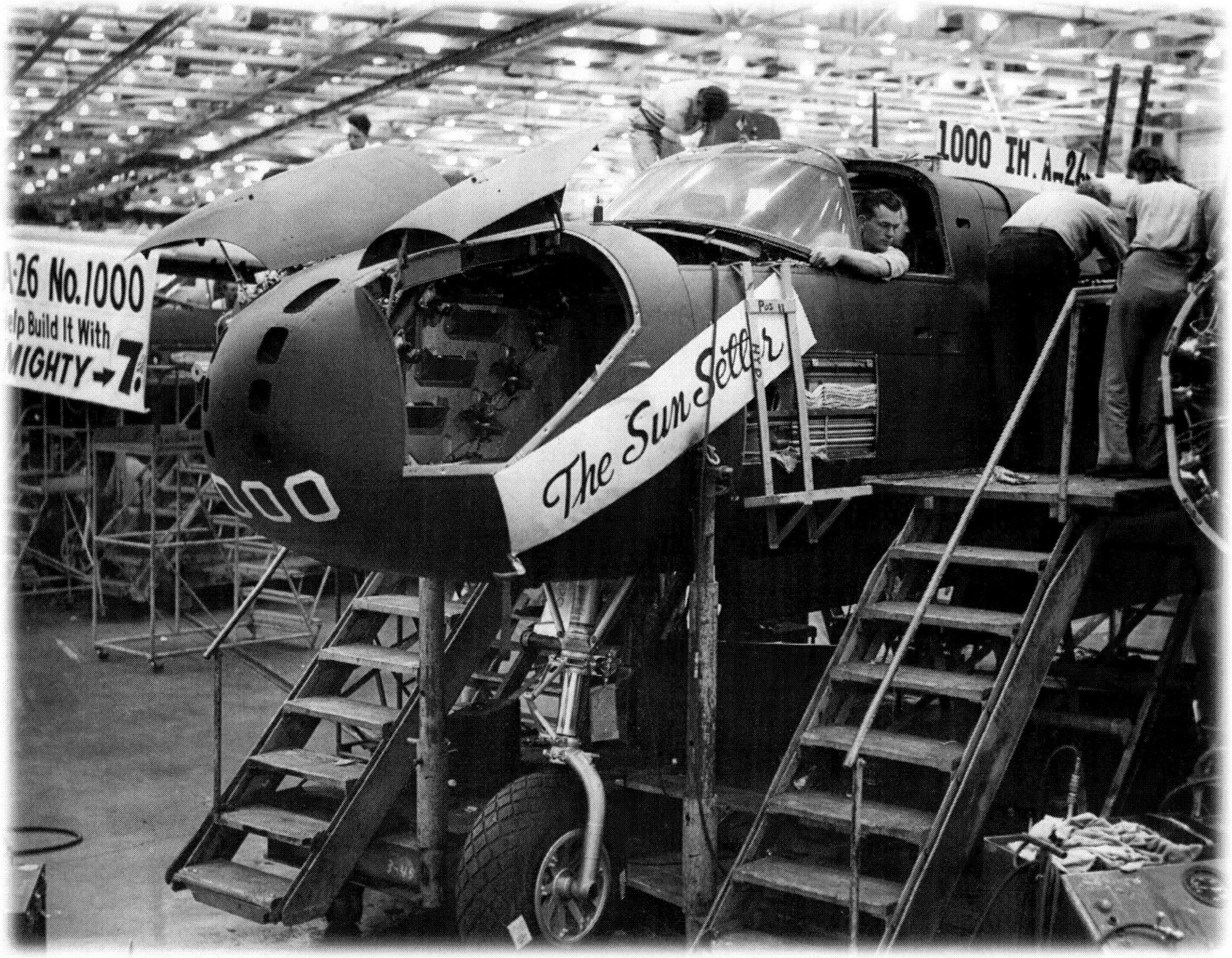

A-26 Invader – This was the 1,000th A-26 aircraft built at Douglas Aircraft. A descendant of the A-20, the initial A-26s were designed as a single-pilot aircraft. In the B designation, alongside the pilot, a crew member typically served as navigator and gun loader for the pilot-operated nose guns. There were a lot of nose guns too; the solid nose version housed six or eight .50 caliber machine guns. In addition to the nose guns, some versions of the aircraft had three .50 caliber machine guns in each wing plus eight nose guns. It was a formidable weapon. (Douglas Aircraft Company)

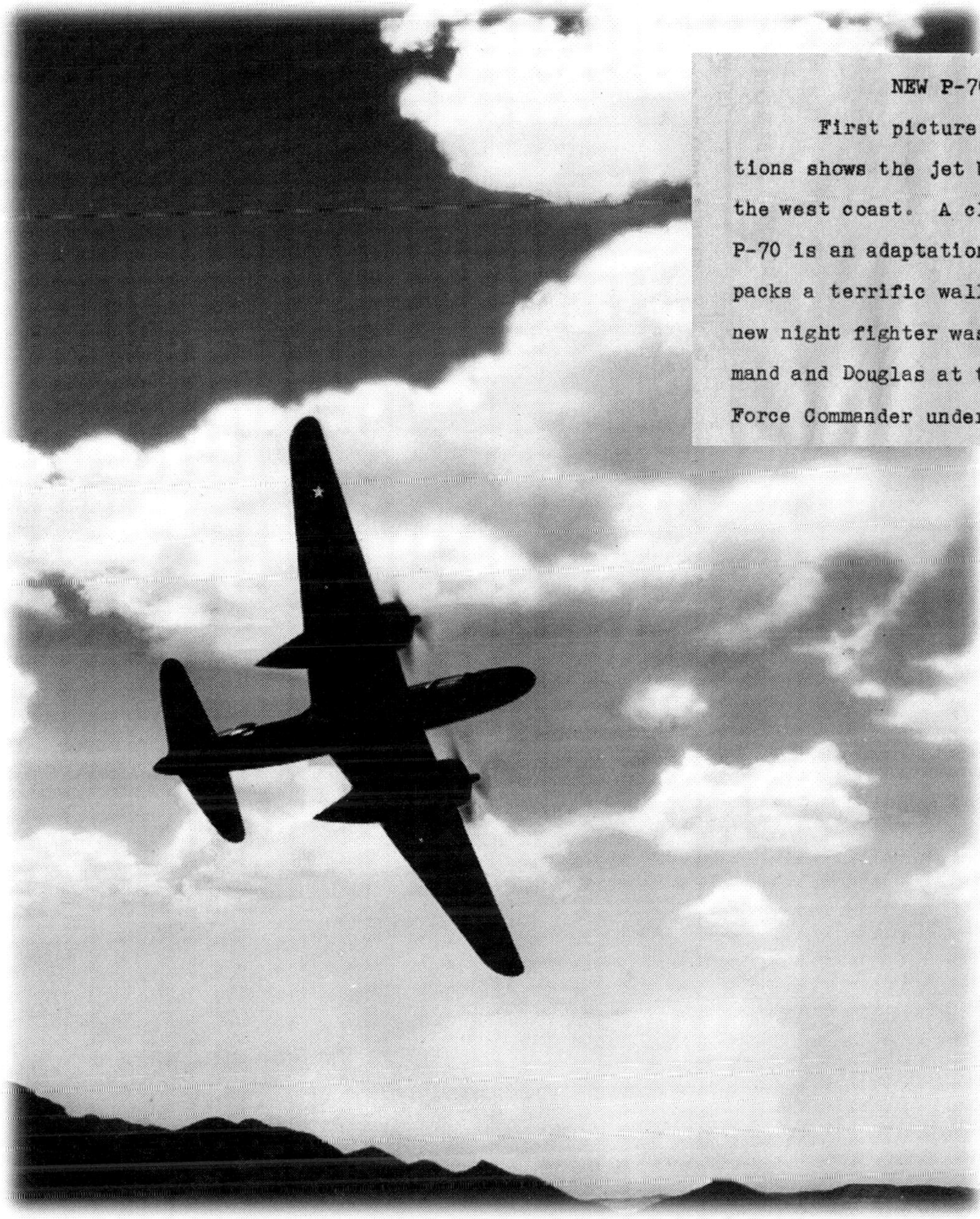

NEW P-70 NIGHT FIGHTER IS DEADLY PROWLER

First picture of the mysterious night wrecker of Axis installations shows the jet black and deadly Douglas P-70 on the flight over the west coast. A closely guarded secret for more than a year the P-70 is an adaptation of the famous Douglas A-20 attack-bomber, and packs a terrific wallop in the four 20 mm. cannon in its belly. The new night fighter was developed by engineers of the AAF Materiel Command and Douglas at the request of Lt. Gen. George C. Kenney, Air Force Commander under Gen. Douglas MacArthur. (Original Press Release)

P-70 Nighthawk

Another WWII adaptation of the A-20 bomber was the P-70 Nighthawk. Using the A-20 airframe for the most part, the P-70 had new radars installed in the nose and were used as trainers for a new fighter, the P-61.

A few aircraft made it into combat. The 6th Night Fighter Squadron received its first P-70 aircraft in September 1942 and operated them from Hawaii. In early 1943, two detachments moved to Guadalcanal and New Guinea, taking the P-70 into combat operations. (Douglas Aircraft Company)

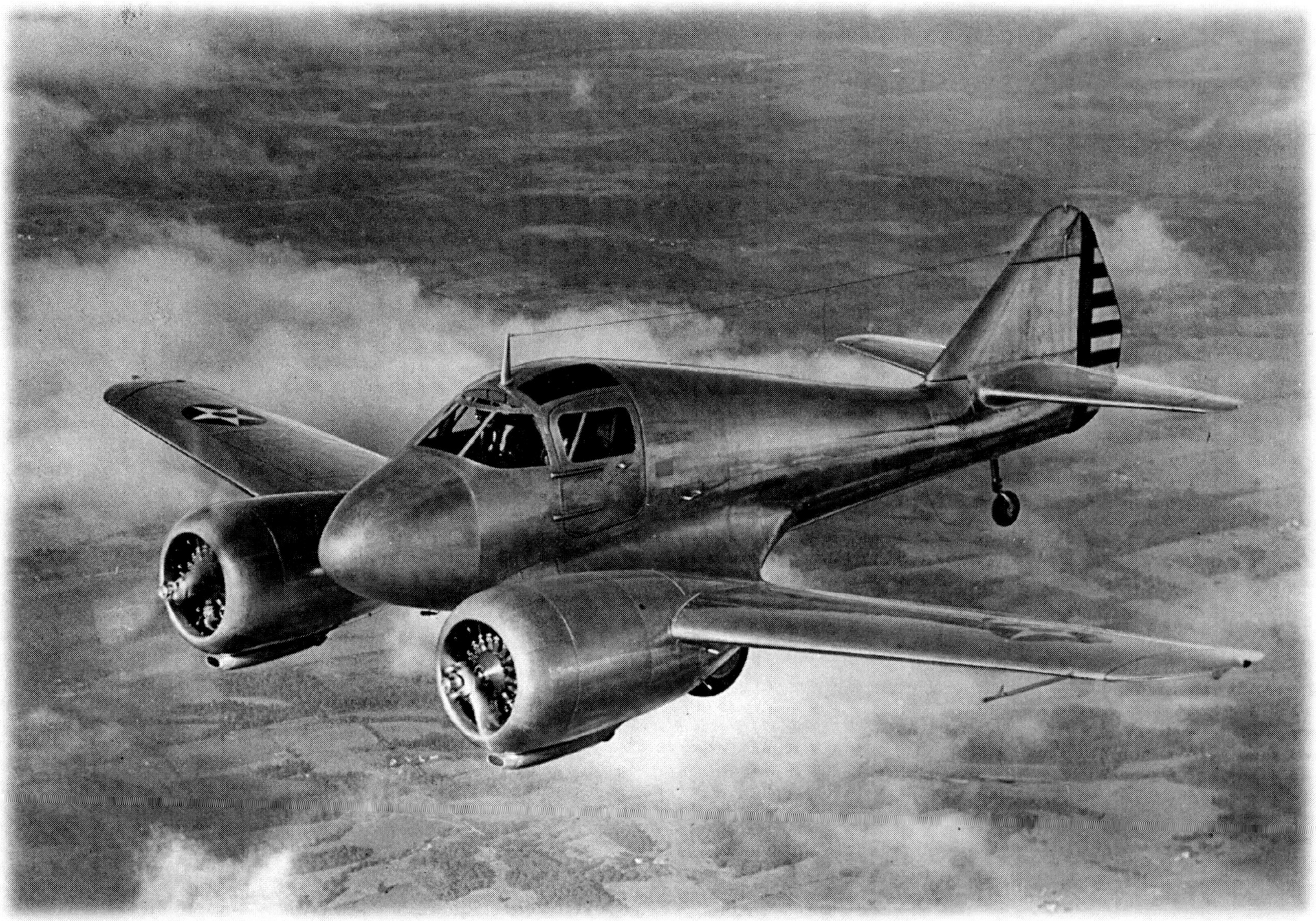

Curtiss-Wright AT-9 – The Curtiss-Wright AT-9 was known as the "Jeep" by its student fliers. It was one of the 1942 trainers for U.S. Army multi-engine pilots. About 800 where produced in two models. (Curtiss-Wright)

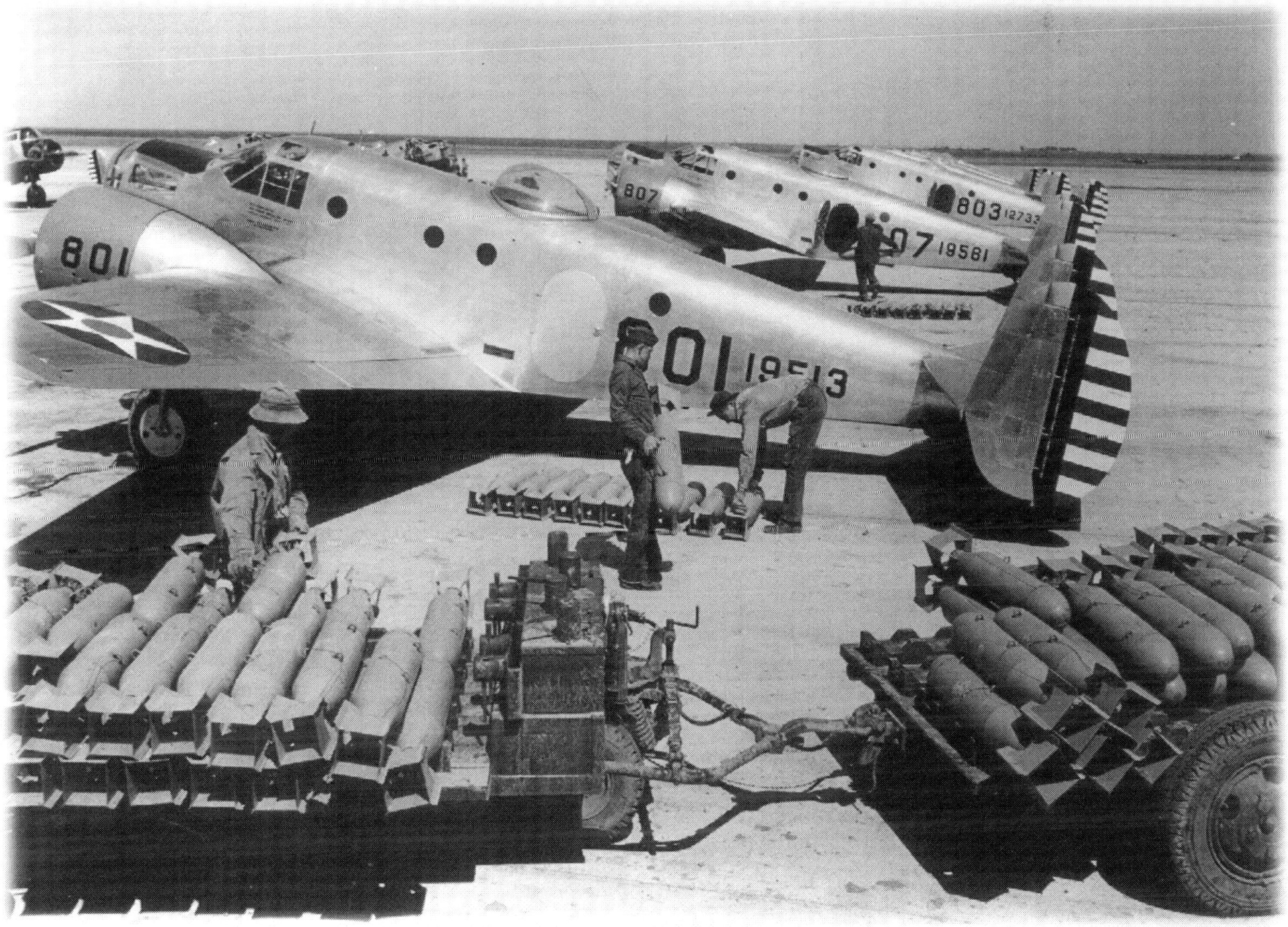

AT-11 Kansan Trainer – Bombardiers and navigators needed experience before they faced the challenges of flying larger aircraft in formation and combat conditions. The AT-11 served as a trainer and carried ten 100 pound practice or general purpose high explosive bombs. The typical crew consisted of a pilot, the instructor and two students. Each student would take a turn in the nose behind the Norden or Sperry bombsight as the instructor looked on from his right side. Each student would drop one bomb per run for five separate bombing runs and would then trade places with the other student. The student who was not at the bombsight would use a 35mm motion picture camera to film each bomb run through a hole in the floor to record the other student's bombing scores.[1] (U.S. Army Air Forces) [1]www.twinbeech.com/at-11page1.htm

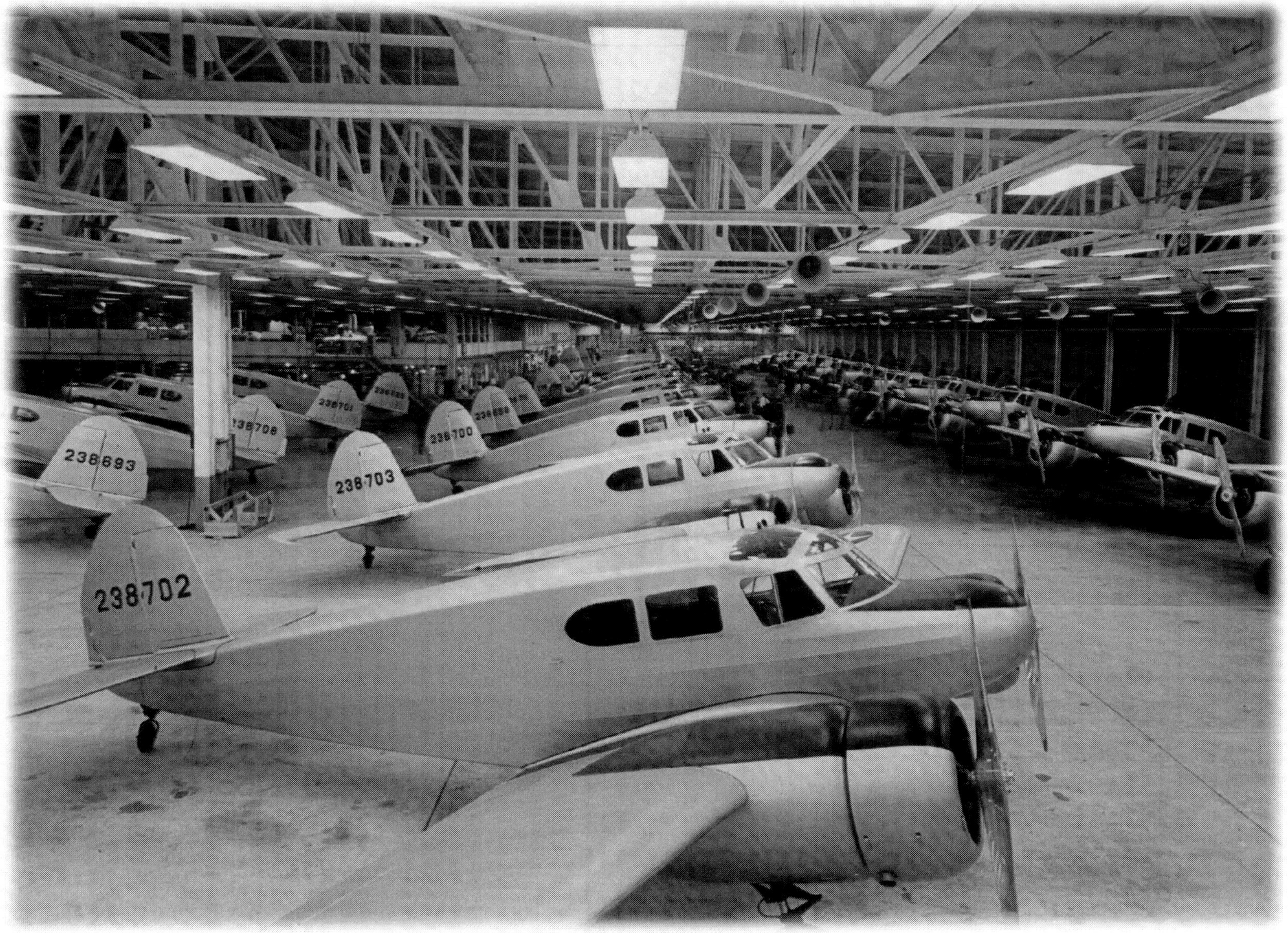

Cessna AT-8 – The Cessna AT-8 went by several names, one of the most popular was Bobcat. Bobcats were used to train bomber pilots for the big B-24, B-17 and other multi-engine bombers used in Europe and the Far East. The planes rolled off the assembly line 24 hours a day and delivered multi-engine experience to thousands of pilots. (Cessna)

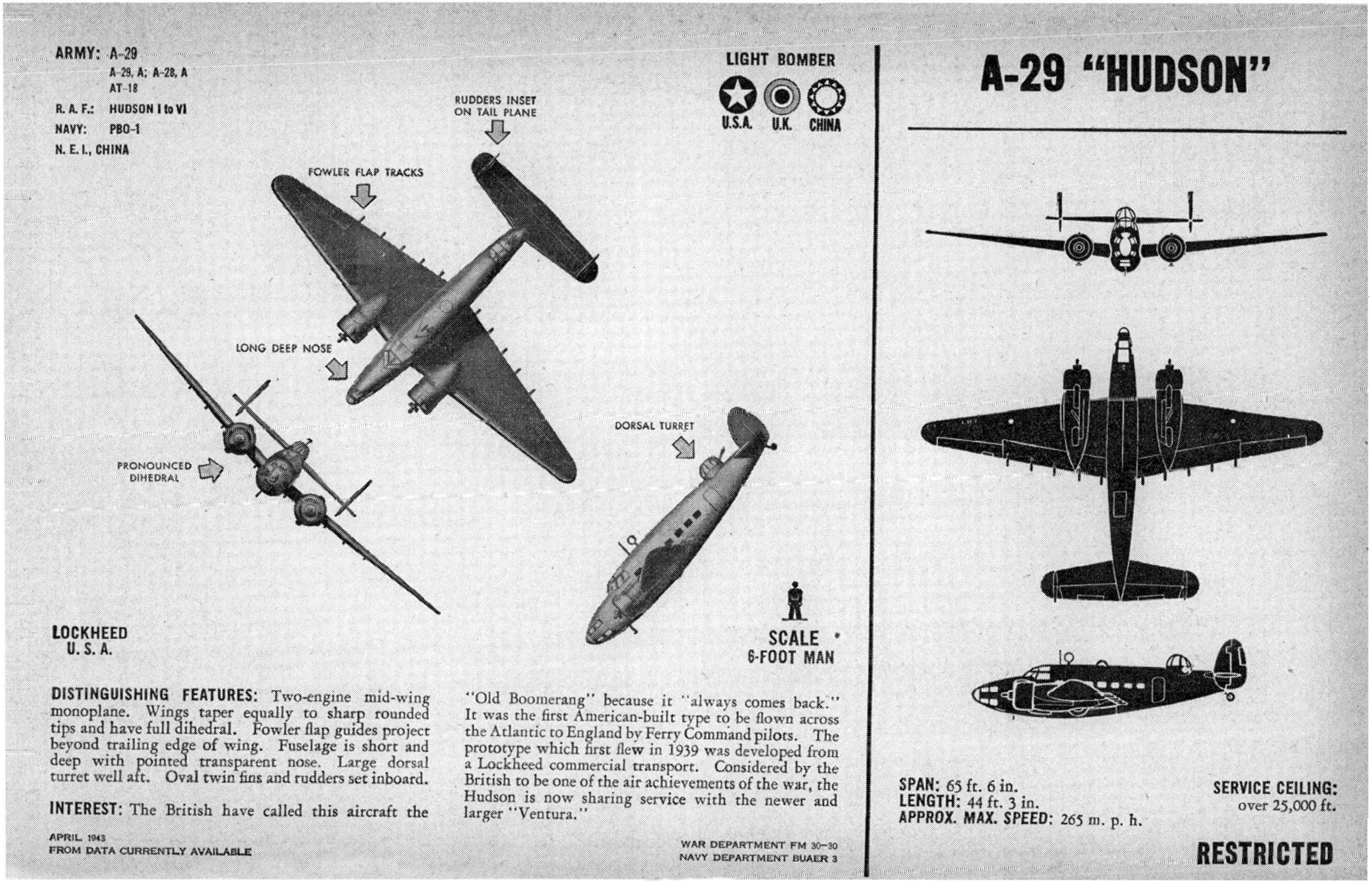

ARMY: A-29
A-29, A; A-28, A
AT-18
R. A. F.: HUDSON I to VI
NAVY: PBO-1
N. E. I., CHINA

LIGHT BOMBER

U.S.A. U.K. CHINA

A-29 "HUDSON"

LOCKHEED
U. S. A.

SCALE • 6-FOOT MAN

DISTINGUISHING FEATURES: Two-engine mid-wing monoplane. Wings taper equally to sharp rounded tips and have full dihedral. Fowler flap guides project beyond trailing edge of wing. Fuselage is short and deep with pointed transparent nose. Large dorsal turret well aft. Oval twin fins and rudders set inboard.

INTEREST: The British have called this aircraft the "Old Boomerang" because it "always comes back." It was the first American-built type to be flown across the Atlantic to England by Ferry Command pilots. The prototype which first flew in 1939 was developed from a Lockheed commercial transport. Considered by the British to be one of the air achievements of the war, the Hudson is now sharing service with the newer and larger "Ventura."

SPAN: 65 ft. 6 in.
LENGTH: 44 ft. 3 in.
APPROX. MAX. SPEED: 265 m. p. h.

SERVICE CEILING: over 25,000 ft.

APRIL 1943
FROM DATA CURRENTLY AVAILABLE

WAR DEPARTMENT FM 30-30
NAVY DEPARTMENT BUAER 3

RESTRICTED

A-29 Hudson – The A-29 Hudson was built by Lockheed and was one of the first operational American aircraft to be used by the RAF during WWII. It filled the British requirement for a maritime patrol/navigational trainer aircraft. Similar to the A-28 bomber, the primary difference was the engine selection. Put yourself into the shoes of a new recruit as you look at these recognition pages. Imagine one of these aircraft silhouettes flashing in front of you along with enemy aircraft silhouettes. You need to make the decision, friend or foe, and give the order to shoot or withhold firing a weapon. (War Department Recognition Guide 1943)

A-29 Hudson – The U.S. Army Air Corps acquired 153 A-29s early in WWII for U.S. coastal defense and antisubmarine patrol squadrons. At least one submarine was sunk in the summer of 1942 by a U.S. A-29. The U.S. Navy also used a small number of A-29s in patrol bomber missions and re-designated them as PBO-1.[1]

(British Combine) [1]www.nationalmuseum.af.mil/factsheets/factsheet.asp?id=3159

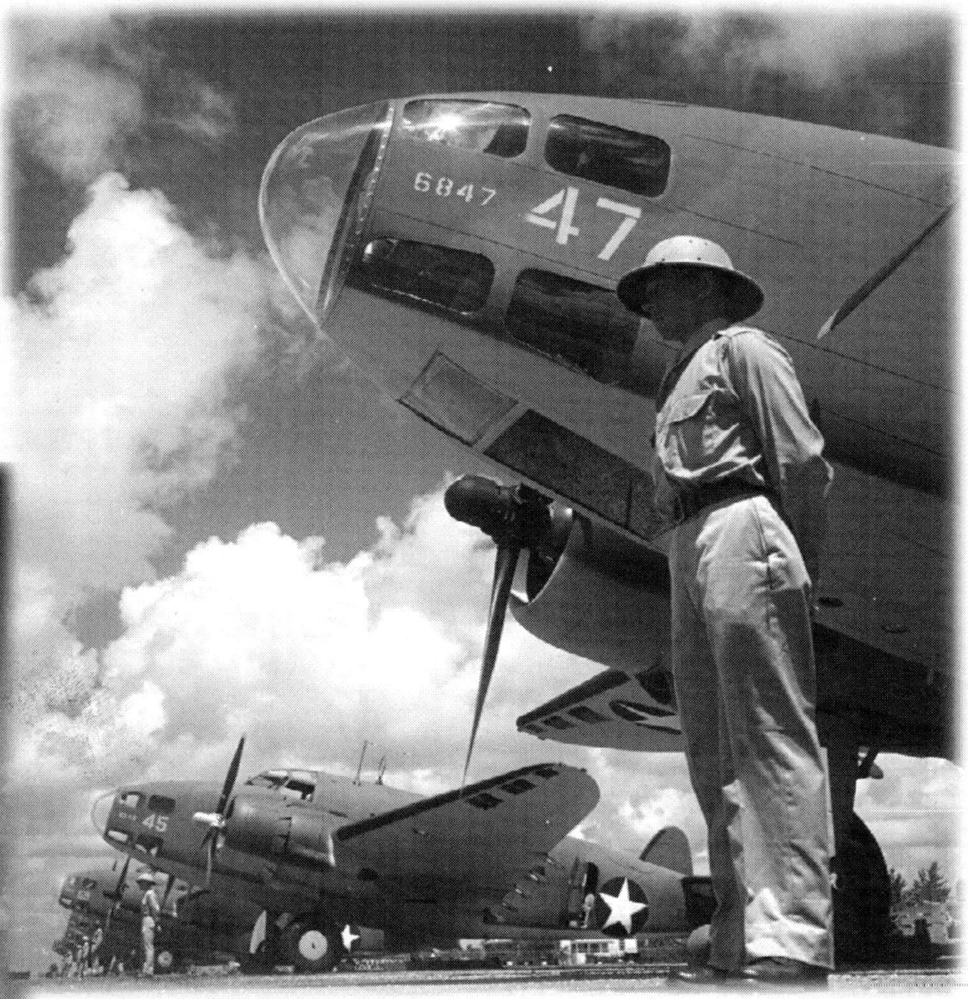

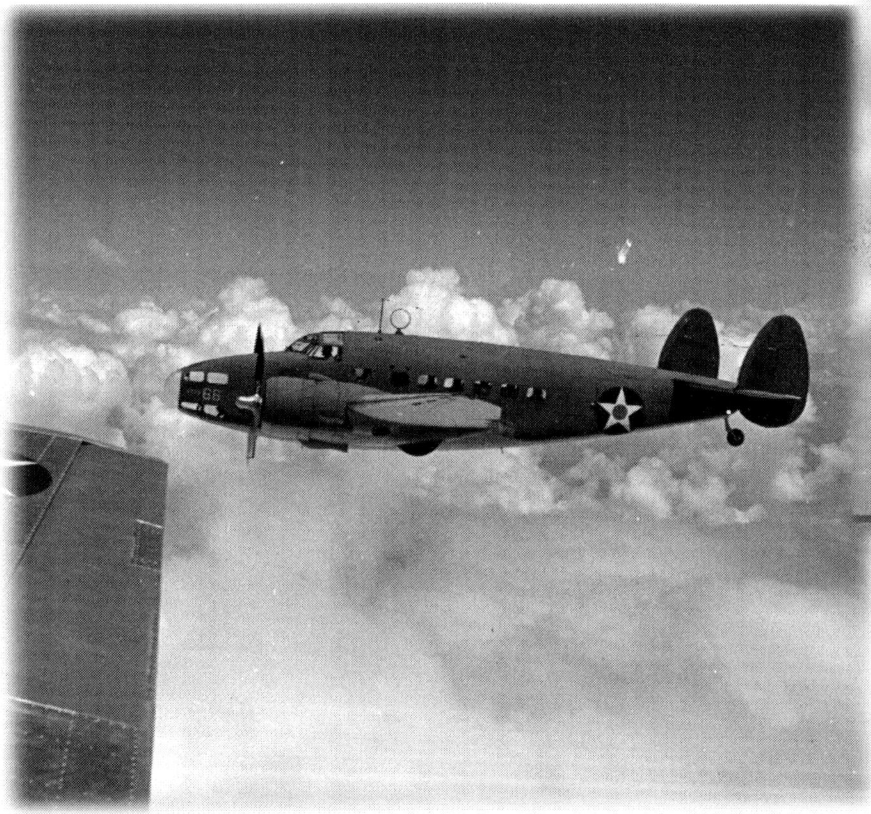

A-29 Hudson – The RAF order for 200 Hudson aircraft was one of the largest orders Lockheed had received prior to the U.S. involvement in WWII. Following the initial Japanese attacks on Malaya, Hudsons became the first aircraft to make an attack in the Pacific War, sinking a Japanese transport ship, the *Awazisan Maru*, off Kota Bharu, an hour before the attack on Pearl Harbor.[1] (British Combine) [1]*rsa.org.nz*

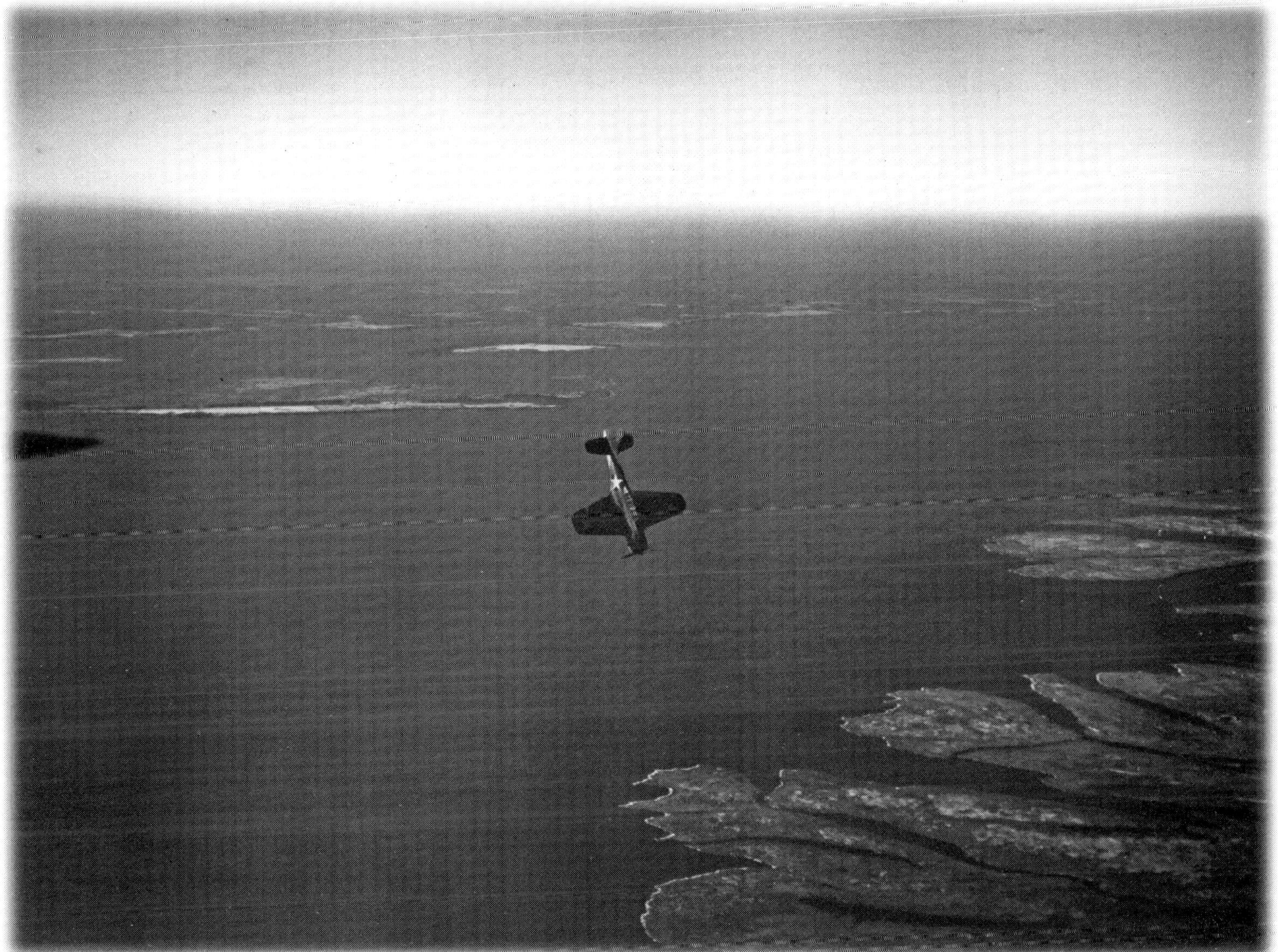

Dive Bombers – Most types of Navy scouting and fighter aircraft also carried out dive-bombing missions. Naval tactics needed to emphasize a relatively small payload because of range and takeoff from carriers. Dive bombers had two features that distinguished them from other combat aircraft. One was the provision for dive brakes allowing the aircraft to slow in the air quickly. The other unique feature was a special hinged bomb rack, or crutch, mounted under the fuselage, which swung the bomb clear of the propeller arc after it was released.[1] (U.S. Navy) [1]www.historynet.com/curtiss-sb2c-helldiver-the-last-dive-bomber.htm/2

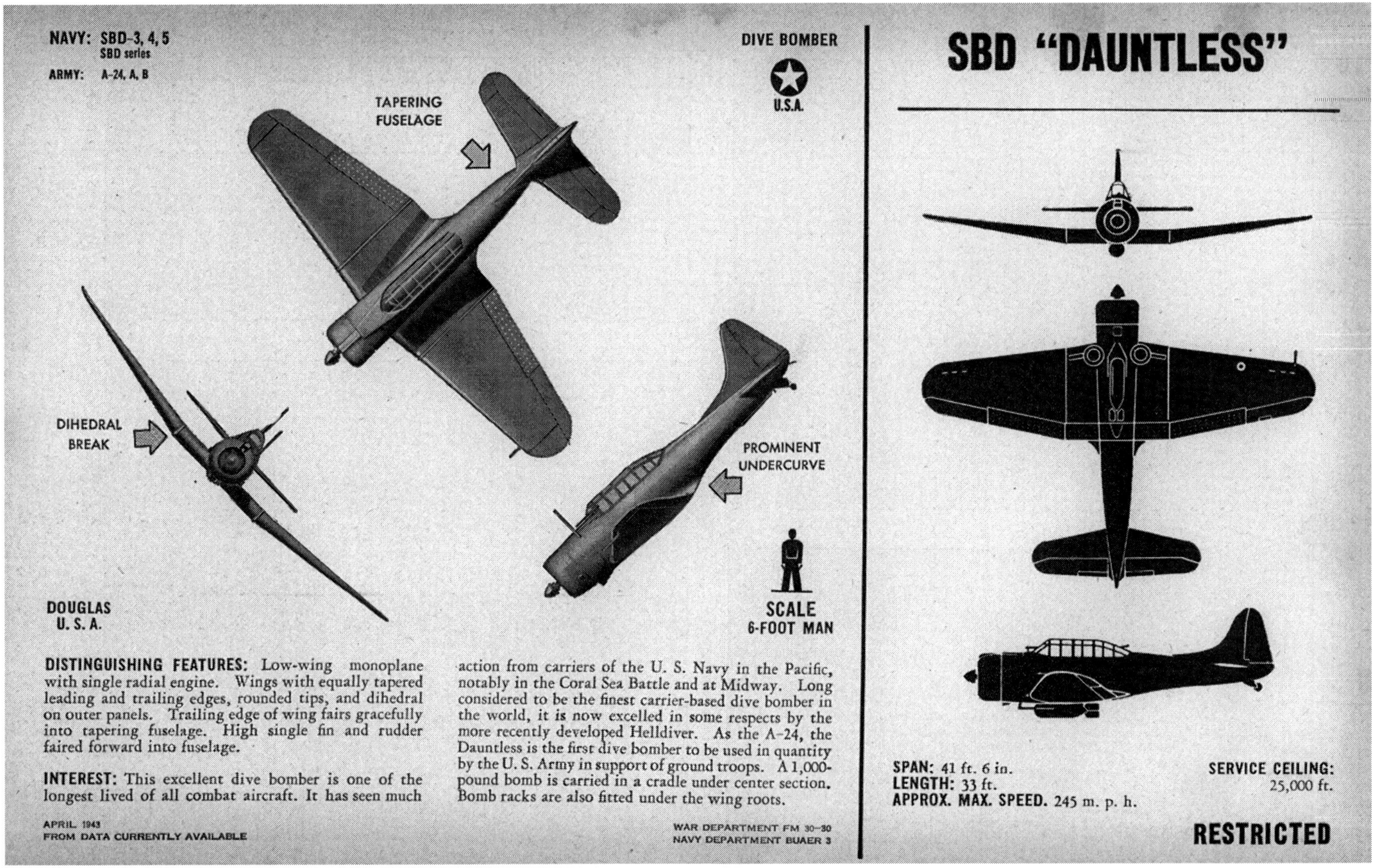

A-24 / SBD Dauntless – The Dauntless was the standard dive bomber at the time of the attack on Pearl Harbor. Designed as a light bomber and reconnaissance aircraft, the Dauntless SBD (Scout Bomber Douglas) monoplanes served during WWII with U.S. Marine Corps and Army and Navy air squadrons. The SBD used perforated flaps and retractable landing gear and was the first U.S. Navy plane to sink an enemy ship in WWII, just three days after Pearl Harbor.[1] (War Department Recognition Guide 1943) [1] Bill Yenne. *McDonnell Douglas, A Tale of Two Giants.* New York: Crescent Books, 1985.

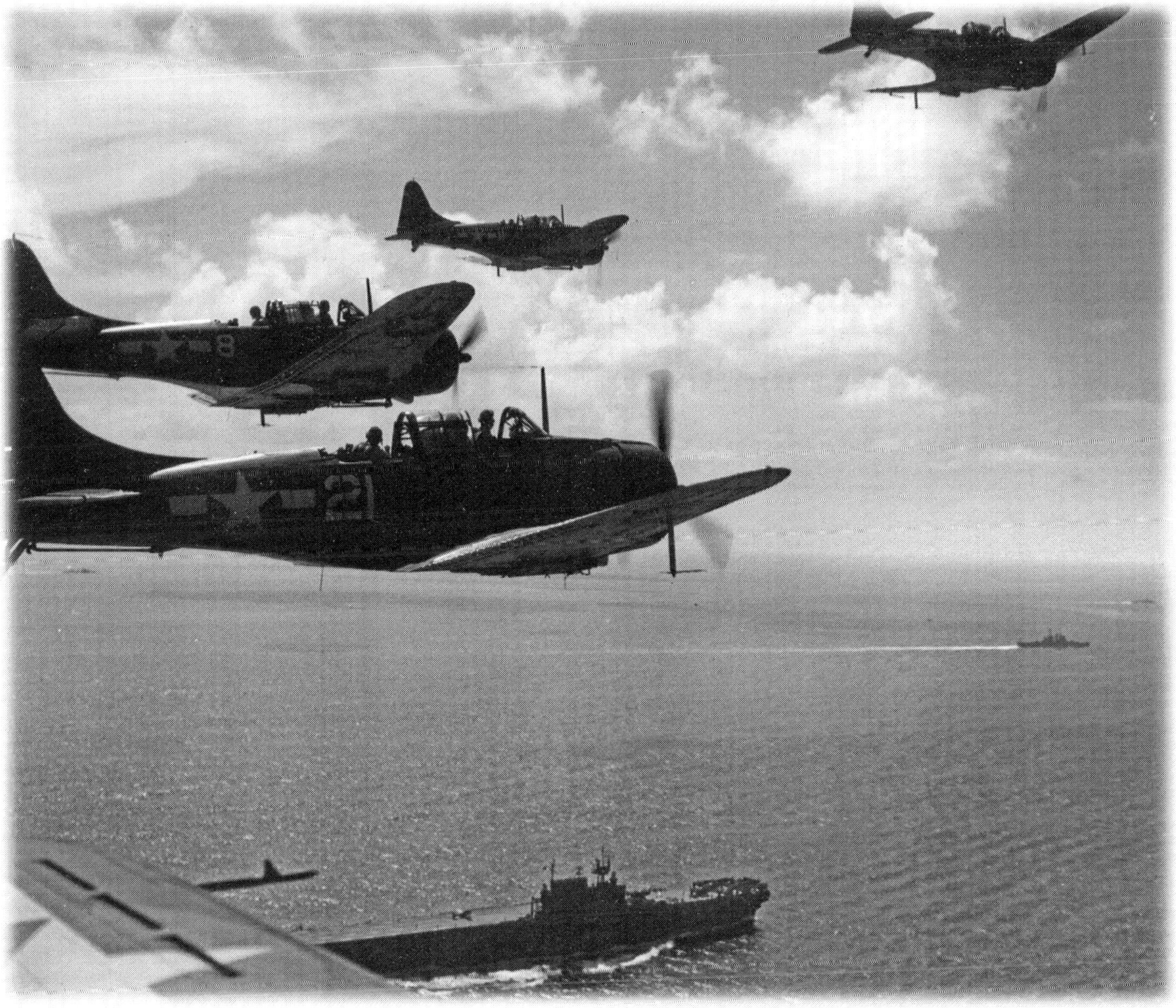

A-24 / SBD Dauntless – This April 1944 photograph shows dive bombers over *Task Force 58* as it roams freely over the Pacific. Dauntless dive bombers sank the Japanese carriers, *Akagi*, *Kaga*, and *Soryu* in the carrier based battle called the Battle of Midway in June 1942. The battle changed the momentum of WWII. (U.S. Navy)

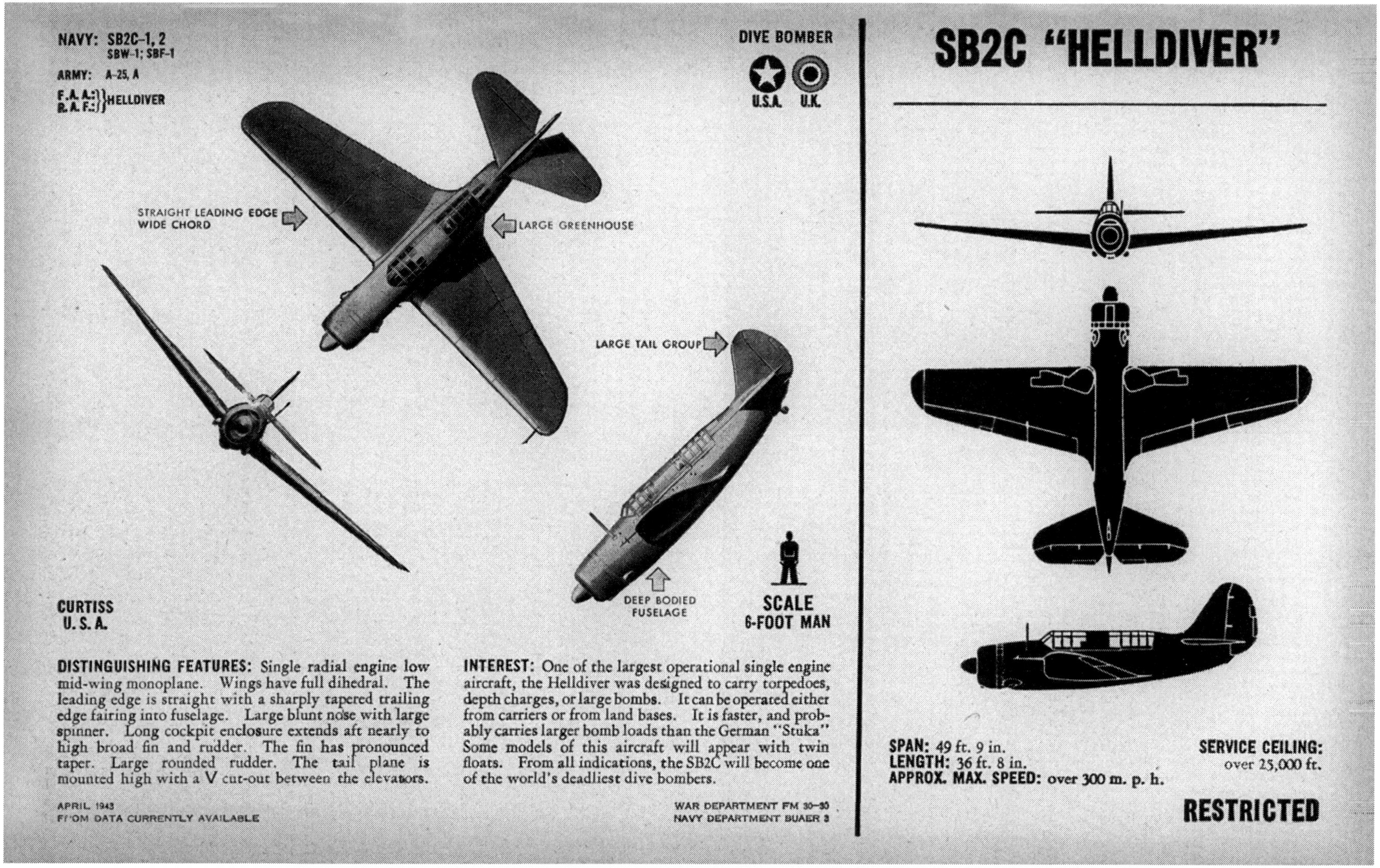

A-25 / SB2C Helldiver – The Curtiss SB2C was the last of a line of aircraft developed for the U.S. Navy specifically for the role of dive bombing.[1] The aircraft was plagued with design issues that delayed its delivery. The Helldiver, also designated A-25, first saw combat with bombing squadron VB-17 from the carrier *Bunker Hill* during an attack on Rabaul on November 11, 1943, nearly three years after the first flight of the prototype.[2] The Helldiver did not exceed the fundamental flight characteristics of the Dauntless, the aircraft it was designed to replace. (War Department Recognition Guide 1943) [1&2]www.historynet.com/curtiss-sb2c-helldiver-the-last-dive-bomber.htm/4

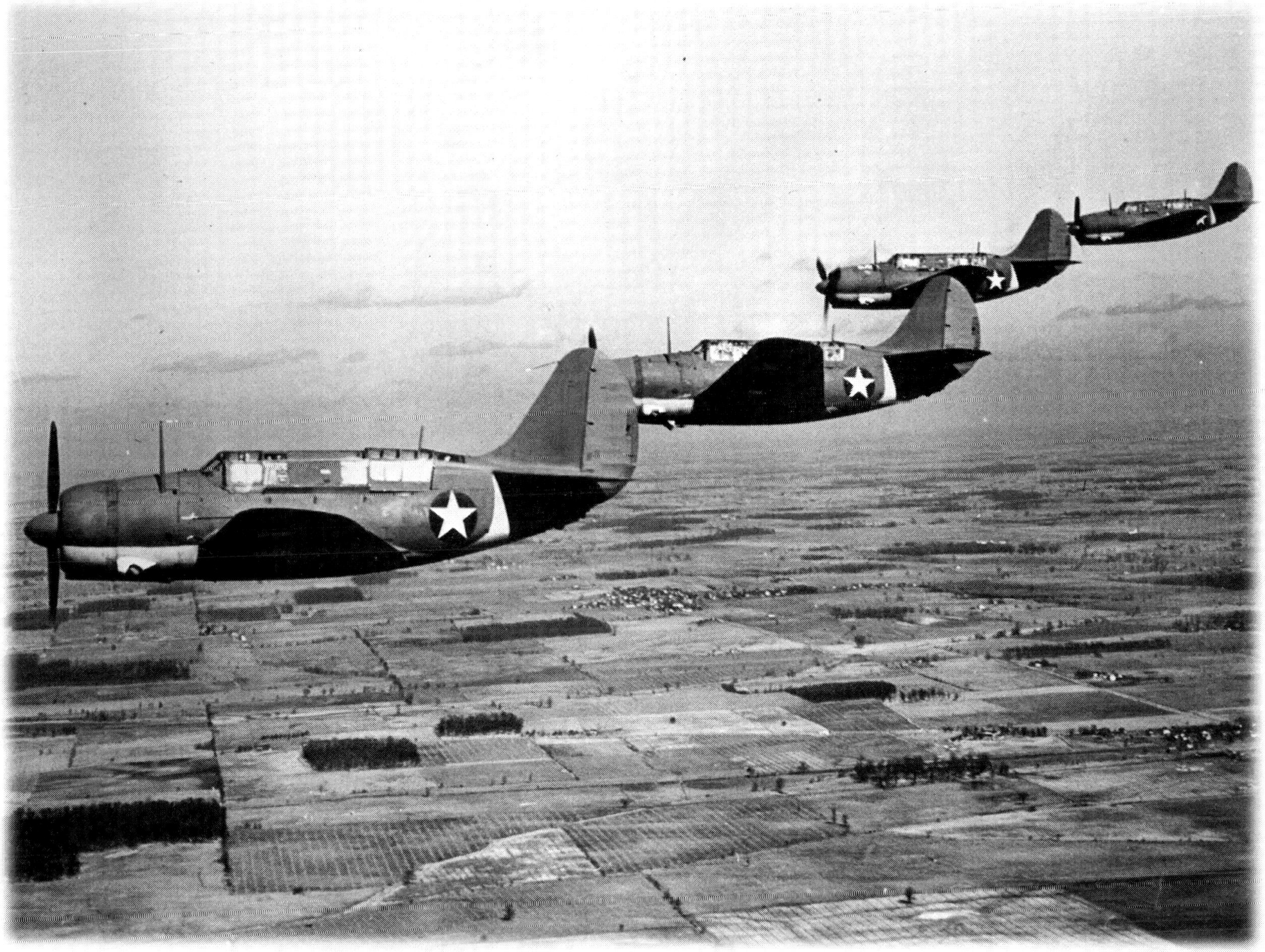

A-25 / SB2C Helldiver – The SB2C took so long in design that when it was finally ready for operational combat, it was no longer needed. The Truman Committee investigated Helldiver production and turned in a scathing report. In spite of the problems, highly trained crews flew the aircraft through the last two years of the Pacific War with a fine combat record. (Curtiss-Wright)

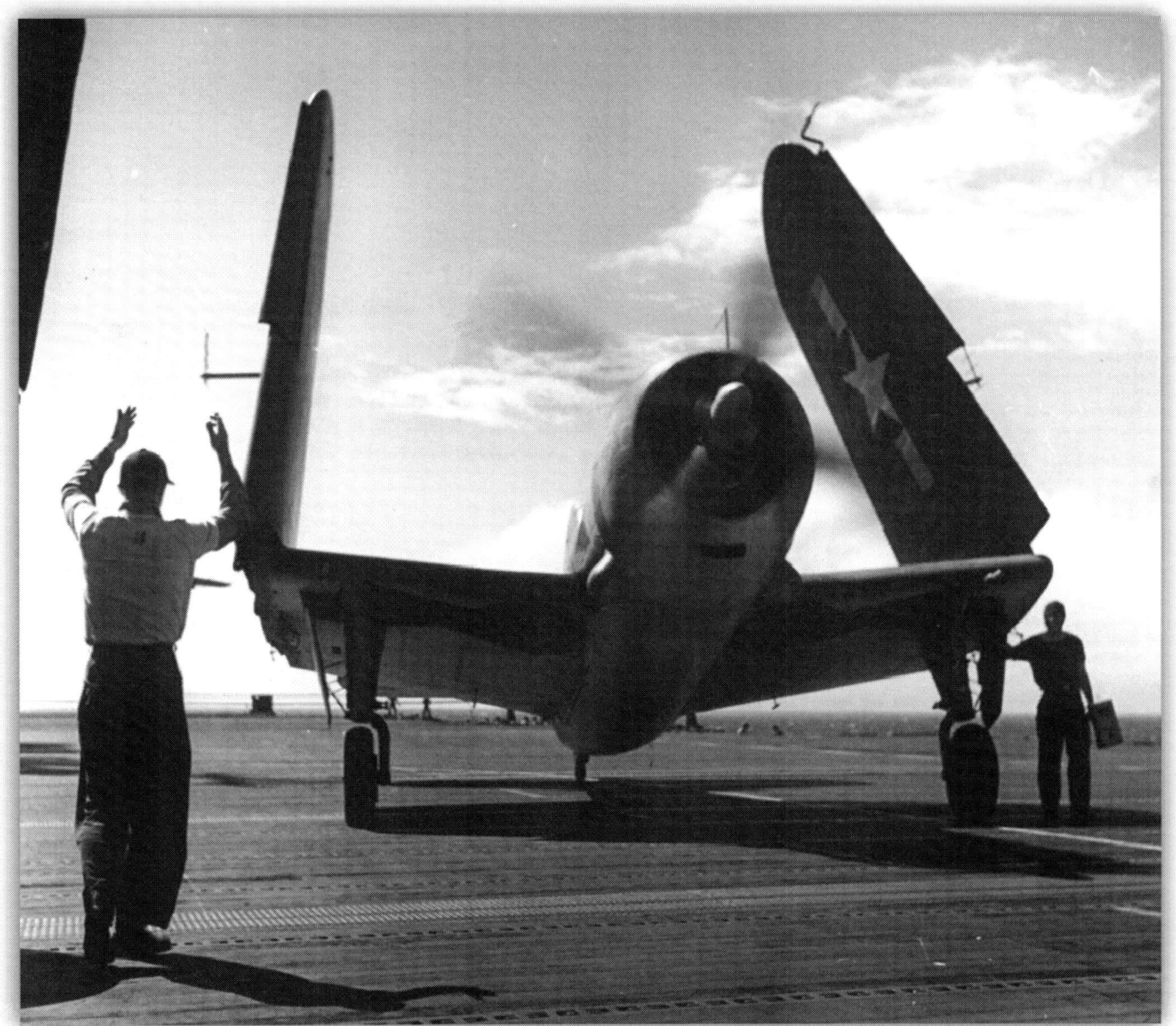

A-25 / SB2C Helldiver – This SB2C rolls forward on the flight deck of an Essex class carrier to be spotted and parked so other aircraft may land aboard. The crewman with his hands over head is a plane director. Notice the sailor beneath the folding port wing with a check block. (U.S. Navy)

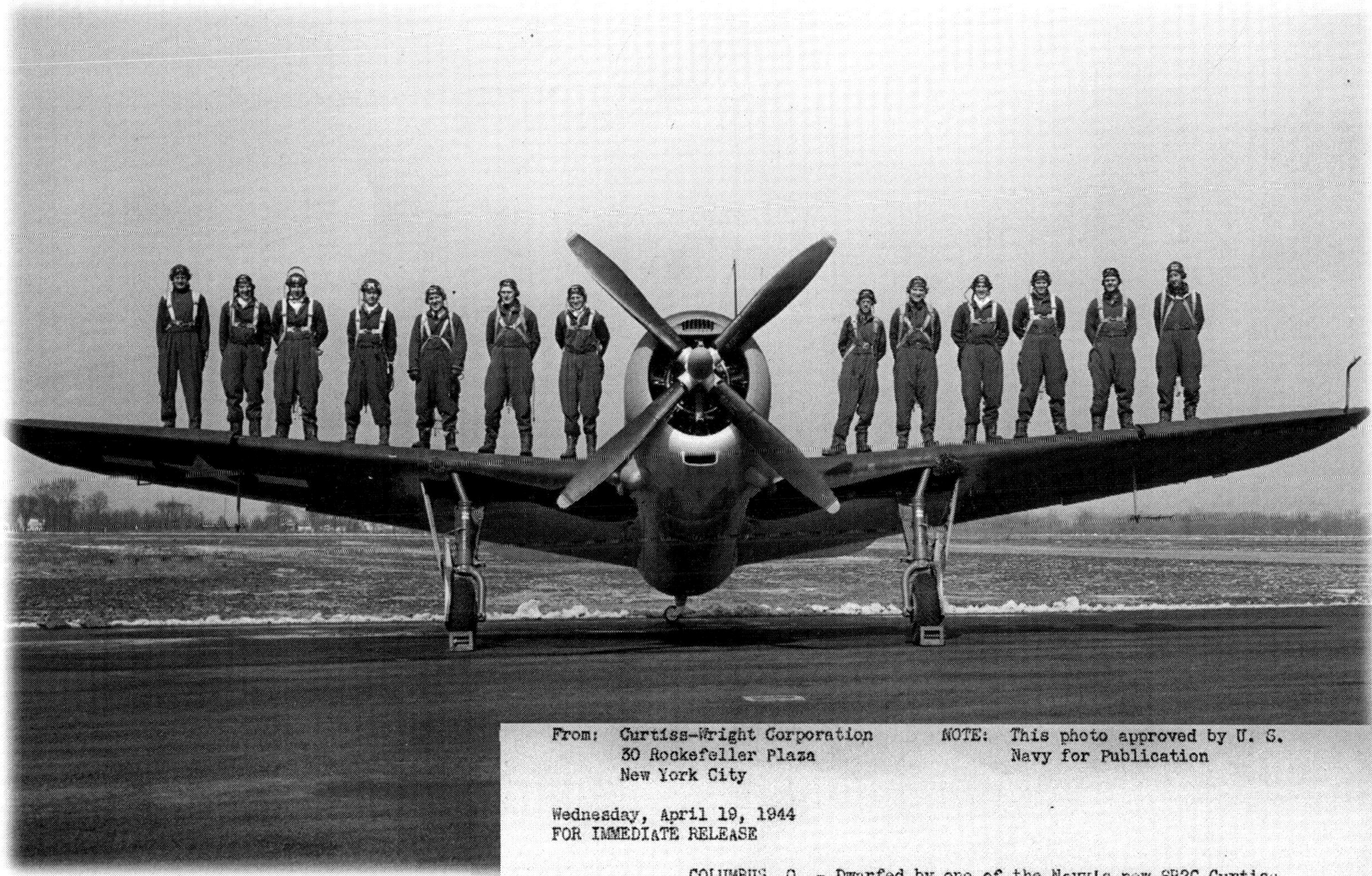

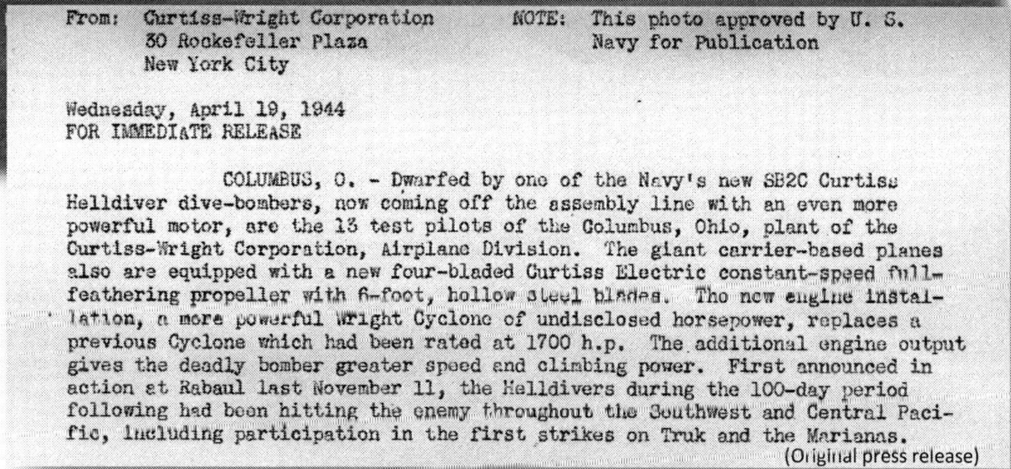

From: Curtiss-Wright Corporation
30 Rockefeller Plaza
New York City

NOTE: This photo approved by U. S. Navy for Publication

Wednesday, April 19, 1944
FOR IMMEDIATE RELEASE

COLUMBUS, O. - Dwarfed by one of the Navy's new SB2C Curtiss Helldiver dive-bombers, now coming off the assembly line with an even more powerful motor, are the 13 test pilots of the Columbus, Ohio, plant of the Curtiss-Wright Corporation, Airplane Division. The giant carrier-based planes also are equipped with a new four-bladed Curtiss Electric constant-speed full-feathering propeller with 6-foot, hollow steel blades. The new engine installation, a more powerful Wright Cyclone of undisclosed horsepower, replaces a previous Cyclone which had been rated at 1700 h.p. The additional engine output gives the deadly bomber greater speed and climbing power. First announced in action at Rabaul last November 11, the Helldivers during the 100-day period following had been hitting the enemy throughout the Southwest and Central Pacific, including participation in the first strikes on Truk and the Marianas.

(Original press release)

A-25 / SB2C Helldiver – Visualizing the size and scale of a Navy dive bomber can be difficult if you have never stood next to one. Photographed in April 1944, thirteen test pilots from the Columbus, OH, Curtiss-Wright plant took a few minutes to demonstrate the size of the aircraft with its new four-bladed propeller. (Curtiss-Wright)

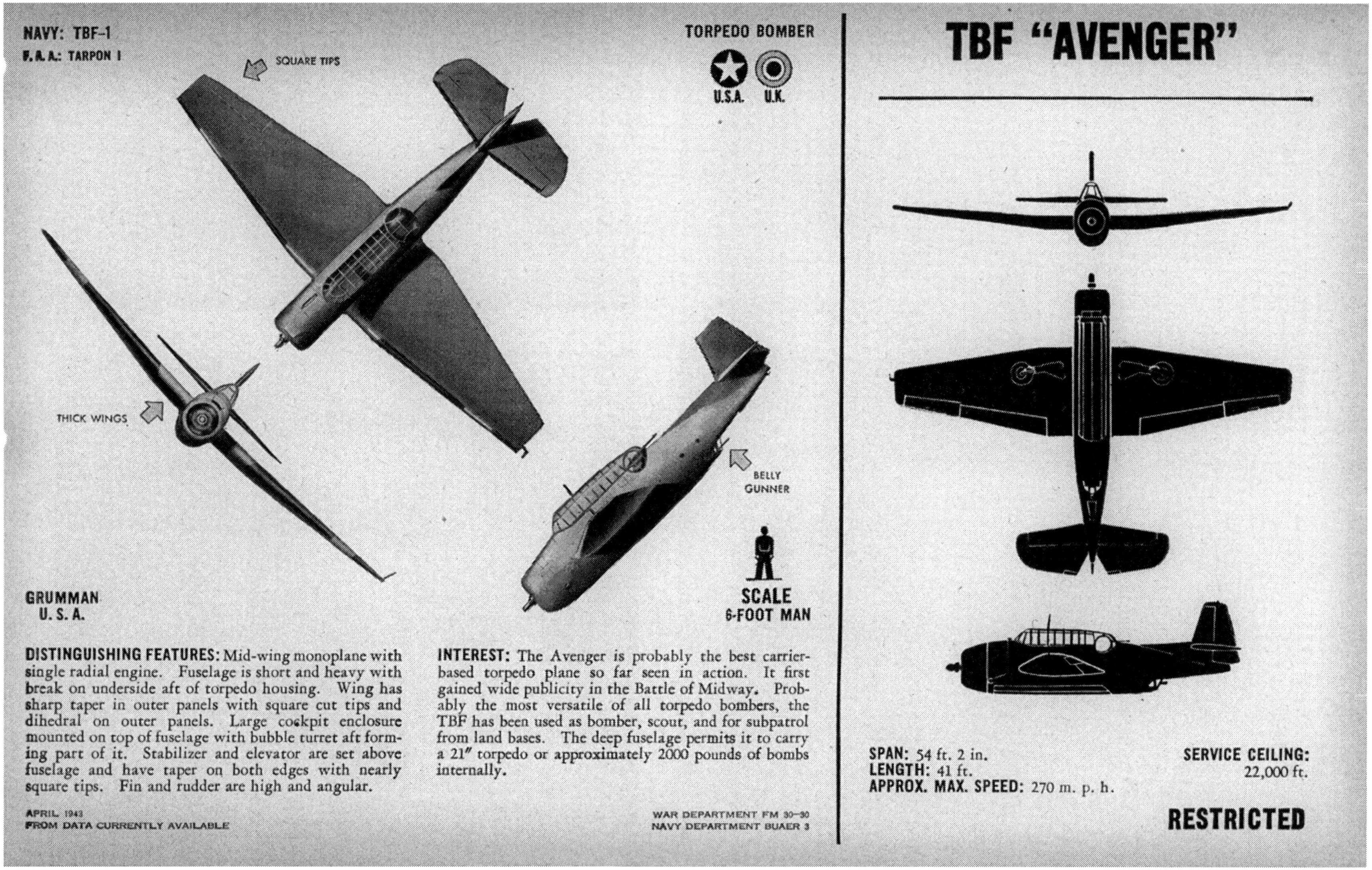

TBF Avenger – The Grumman TBF Avenger torpedo bomber entered U.S. Navy service just in time to participate in the Battle of Midway. Although five of the six aircraft were lost, the aircraft was admired for its durability. As Grumman's first torpedo bomber, it featured a crew of three and became the heaviest single-engine aircraft of WWII. It could carry 2,000 pounds of bombs or one Mark 13 torpedo. Over 9,800 were built during WWII. (War Department Recognition Guide 1943)

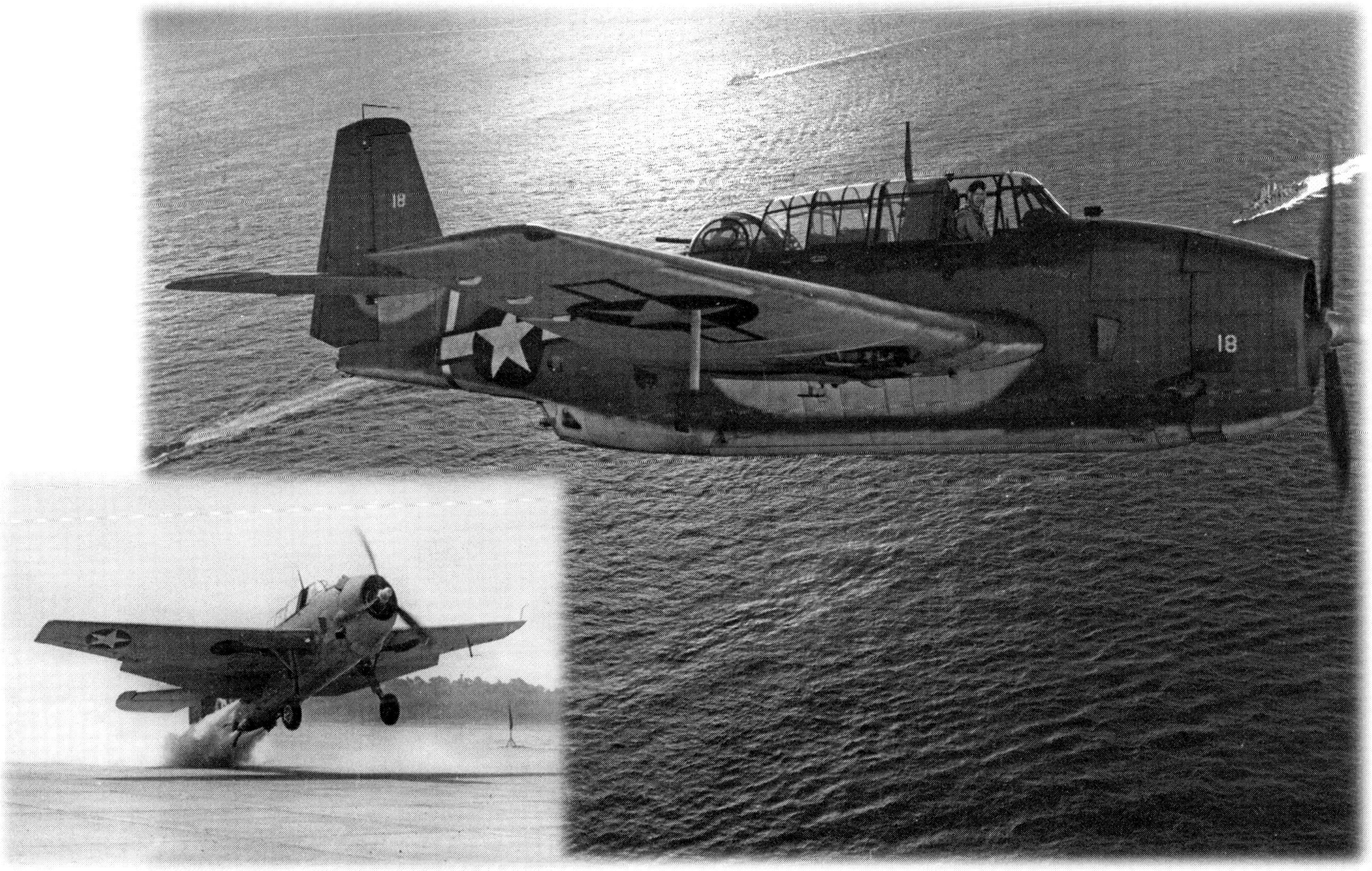

TBF Avenger – The Avenger was well armed in support of its three man crew. One .30 caliber machine gun was mounted in the nose, a .50 caliber (12.7 mm) gun was mounted right next to the turret gunner's head in a rear-facing electrically powered turret, and a single .30 caliber hand-fired machine gun, mounted ventrally (under the tail), was used to defend against enemy fighters attacking from below and to the rear.[1] Later models deleted the nose gun in favor of a .50 caliber gun on each wing. The inset photograph shows a September 1944 view of a JATO jet unit being tested with Avenger. Use of the solid propellant JATO enhancement would shorten takeoffs or allow a larger more heavily-laden aircraft to takeoff. (Both photographs by U.S. Navy)

[1] en.wikipedia.org/wiki/Grumman_TBF_Avenger

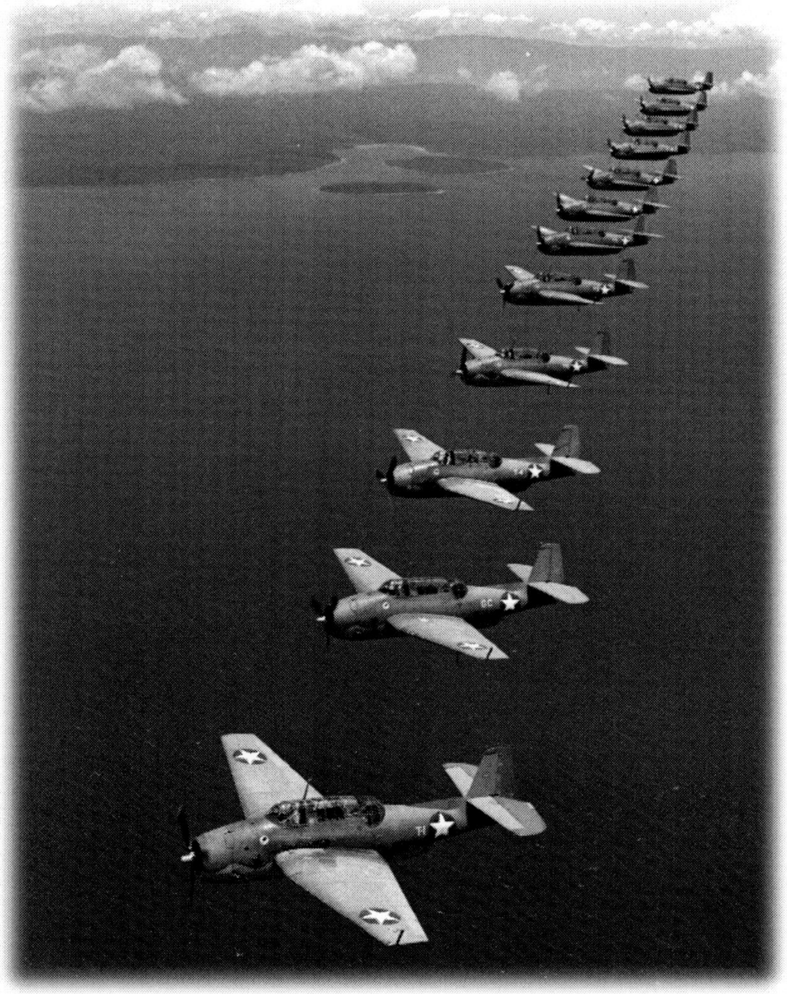

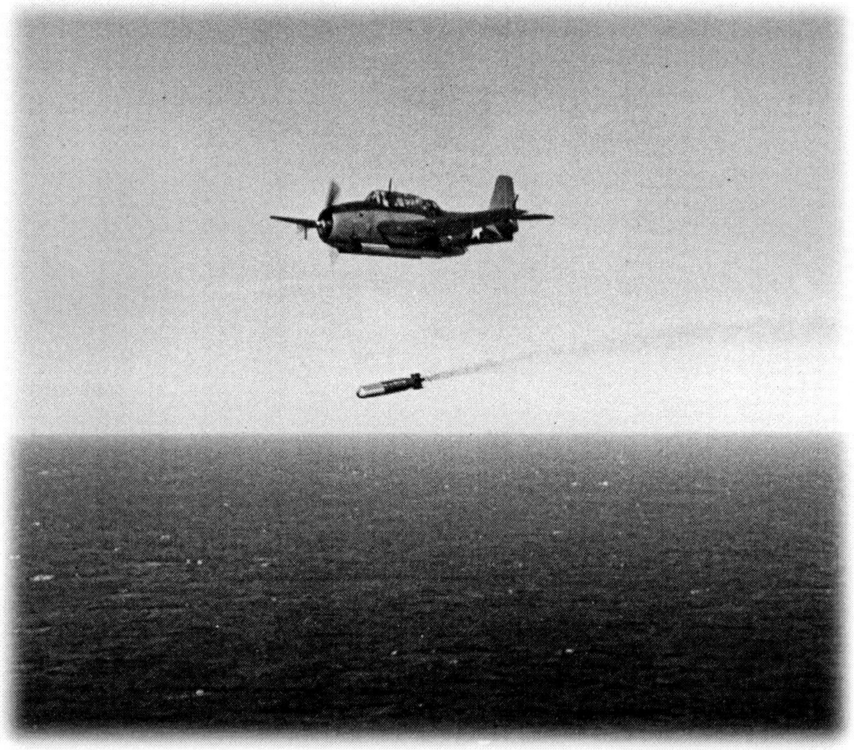

TBF Avenger – The Avenger, like most WWII bombers/dive bombers, flew into battle in a formation. Quantities of aircraft increased the probability that they could deliver maximum damage to the enemy. Large quantities of aircraft were needed to ensure accuracy and that enough explosive was delivered to destroy the target. In the Avenger's case, it could only carry one torpedo.

Many of the Navy aircraft were designed as dive bombers. Other aircraft, like the Corsair, were also effective dive bombers yet were primarily designed and used as a fighter. The Army's P-47 Thunderbolts and P-51 Mustangs were also capable of effective dive bombing yet were formidable fighters. Different aircraft fulfilled unique needs for different types of bombing situations. (Both photographs by U.S. Navy)

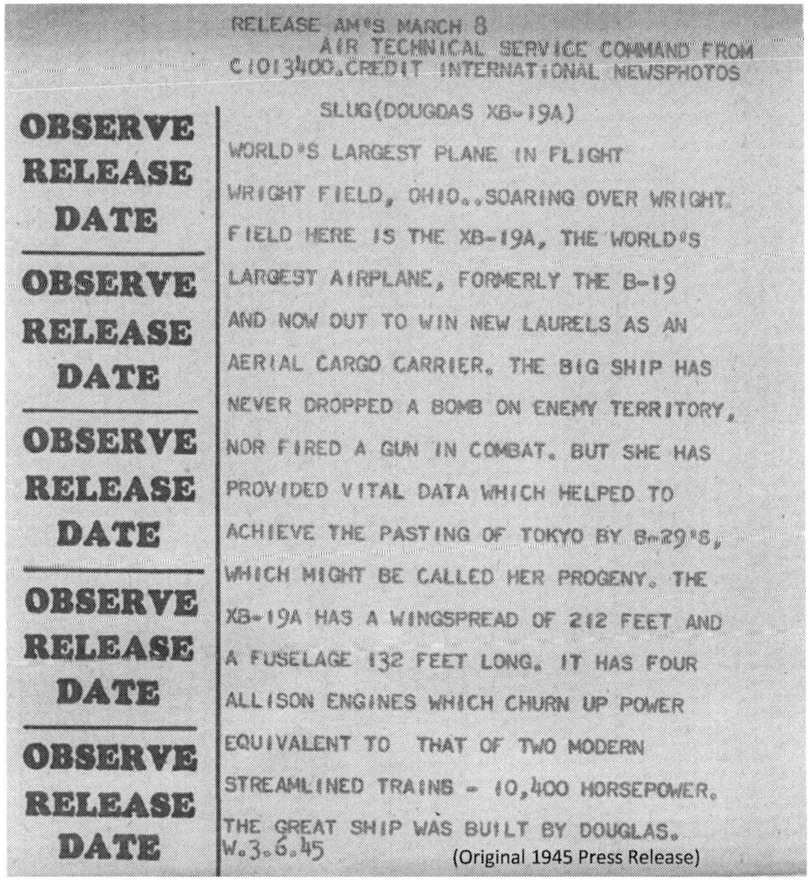

RELEASE AM'S MARCH 8
AIR TECHNICAL SERVICE COMMAND FROM
C1013400.CREDIT INTERNATIONAL NEWSPHOTOS

OBSERVE RELEASE DATE
OBSERVE RELEASE DATE
OBSERVE RELEASE DATE
OBSERVE RELEASE DATE
OBSERVE RELEASE DATE

SLUG(DOUGDAS XB-19A)

WORLD'S LARGEST PLANE IN FLIGHT

WRIGHT FIELD, OHIO..SOARING OVER WRIGHT FIELD HERE IS THE XB-19A, THE WORLD'S LARGEST AIRPLANE, FORMERLY THE B-19 AND NOW OUT TO WIN NEW LAURELS AS AN AERIAL CARGO CARRIER. THE BIG SHIP HAS NEVER DROPPED A BOMB ON ENEMY TERRITORY, NOR FIRED A GUN IN COMBAT. BUT SHE HAS PROVIDED VITAL DATA WHICH HELPED TO ACHIEVE THE PASTING OF TOKYO BY B-29'S, WHICH MIGHT BE CALLED HER PROGENY. THE XB-19A HAS A WINGSPREAD OF 212 FEET AND A FUSELAGE 132 FEET LONG. IT HAS FOUR ALLISON ENGINES WHICH CHURN UP POWER EQUIVALENT TO THAT OF TWO MODERN STREAMLINED TRAINS - 10,400 HORSEPOWER.
THE GREAT SHIP WAS BUILT BY DOUGLAS.
W.3.6.45

(Original 1945 Press Release)

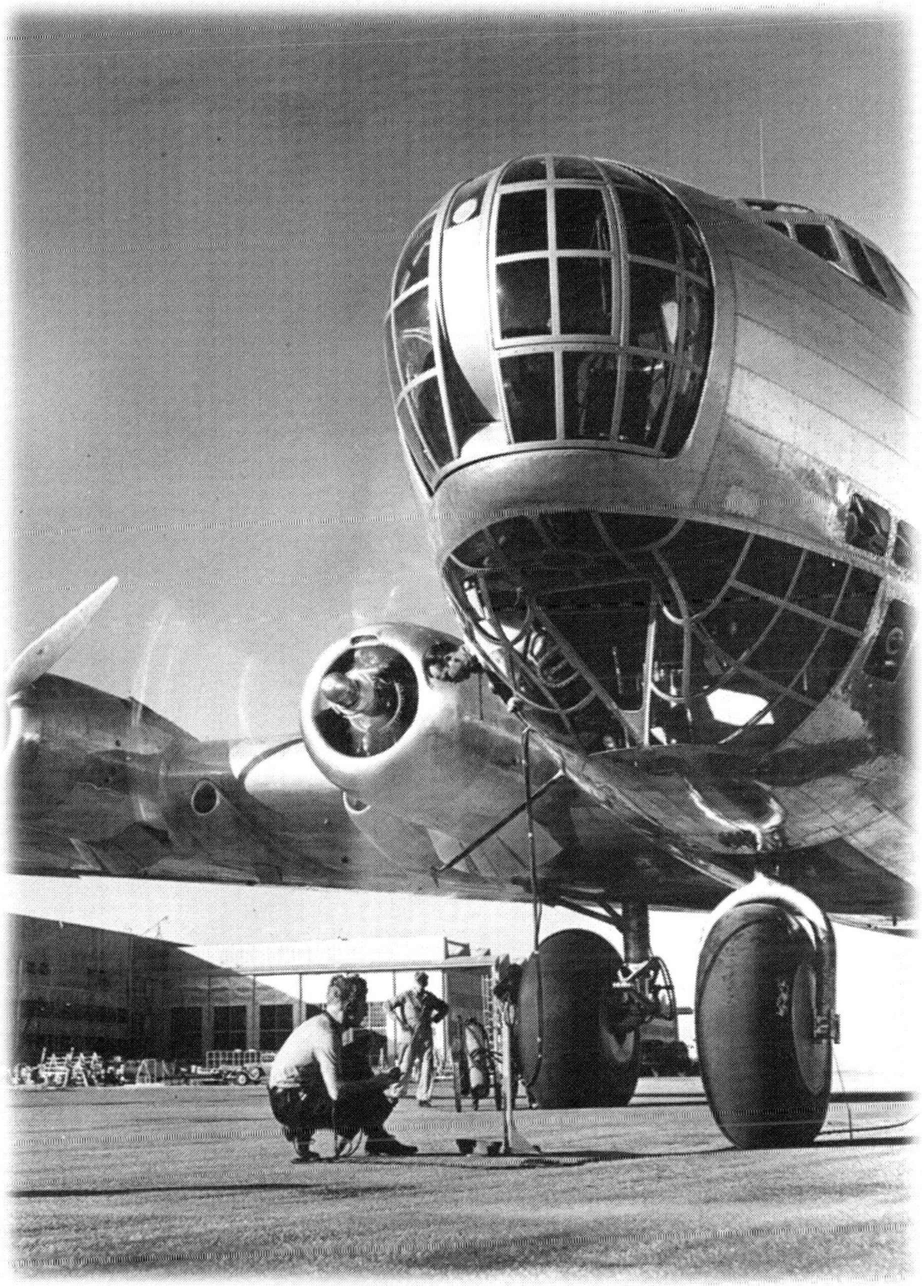

XB-19 – Douglas Aircraft built one XB-19 long range bomber prototype. It was used as a test-bed and an invaluable aid in the design of future Air Force bombers. It was also huge, larger than the production B-29, and it would be the largest WWII combat bomber built. The prototype required a crew of eighteen, and was ordered by the U.S. Army on September 29, 1936. It helped influence the designs of the WWII Boeing B-29 and Cold War Convair B-36. The XB-19 aircraft in the photograph right, provided transport service until 1949 when it was cut up for scrap. (Kaufmann and Fabry Company)

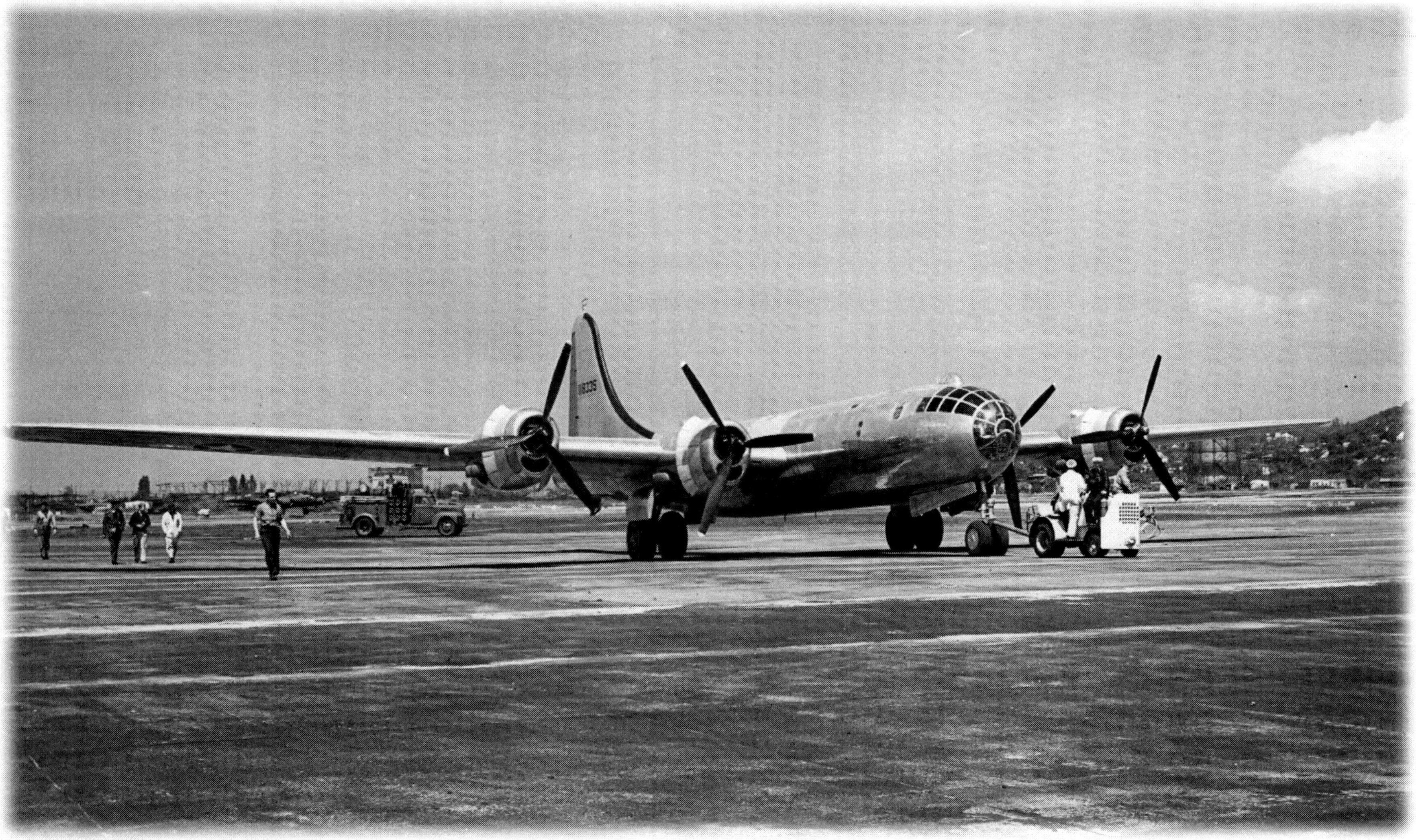

B-29 Superfortress – In the photograph above, an experimental XB-29 bomber was photographed at Boeing in its early developmental stages. The early B-29s used three bladed propellers. Production models later used four bladed Hamilton Standard propellers. The aircraft was 99 feet long and 29 feet high with a wingspan of 141 feet. It weighed 133,500 pounds gross and was powered by four Wright R-3350-23 engines delivering 2,200hp. Before the first B-29 ever flew, 1,650 were on order by the U.S. Government.[1] The first 250 were ordered on May 17, 1941.[2] In 1943, the first production model B-29 rolled out of the Wichita plant. Building the B-29 was crucial because it was the only U.S. bomber with a range of 3,200 miles that could reach Japan and return to base in WWII. (Boeing) [1]*Boeing B-29 Superfortress* by Earl Swinhart [2]*The B-29 Superfortress chronology*, 1934-1960 By Robert A. Mann

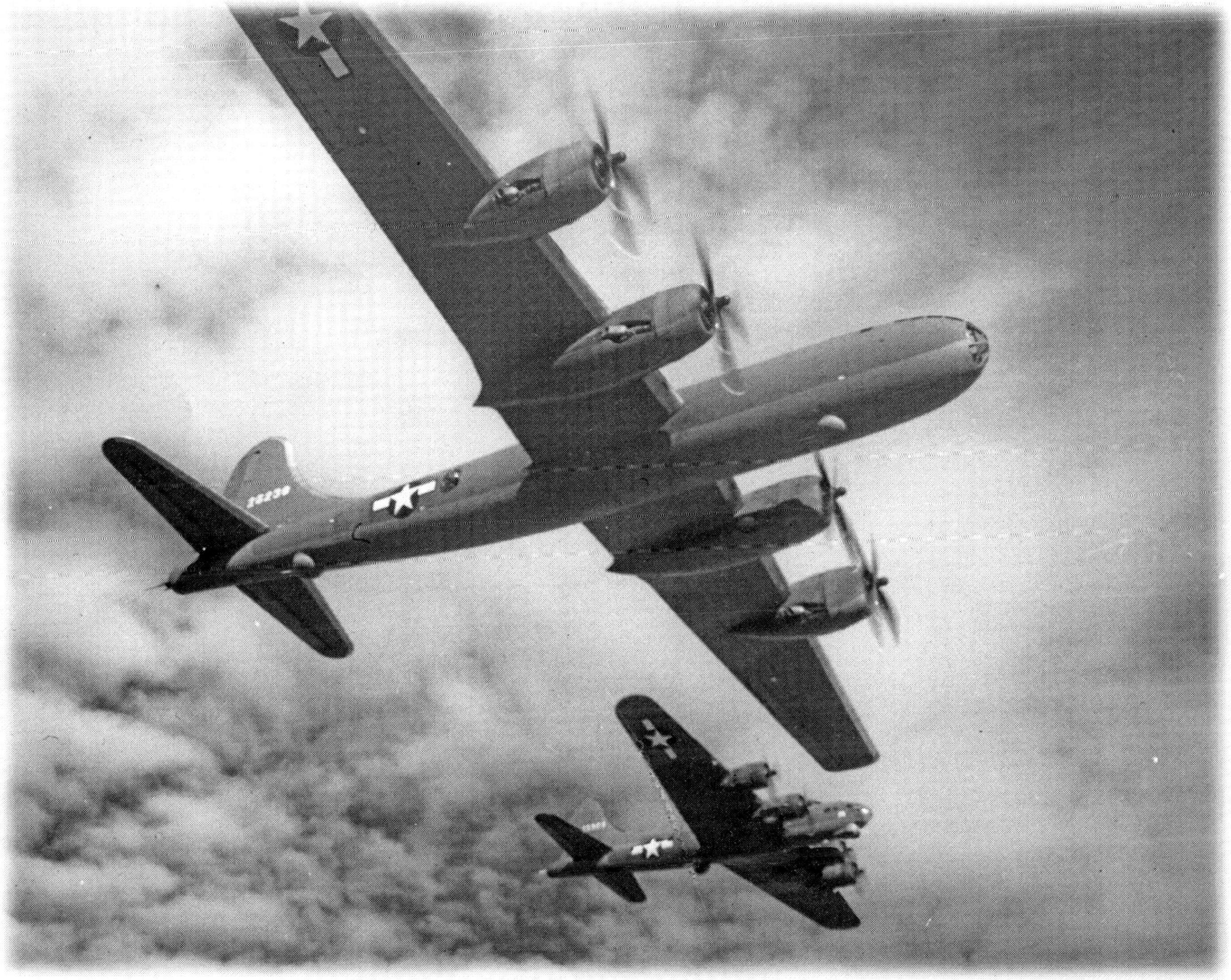

B-29 Superfortress alongside a B-17 – The B-29 was designed for long distance missions and required unique engineering. The aircraft was designed for missions stretching up to 18 hours at altitudes of 32,000 feet. The 141 foot wingspan covered an area of 1736 square feet. By contrast, the B-17, America's initial heavy bomber in WWII, had a wingspan a little longer than 103 feet, covering 1420 square feet. The B-29 was the first pressurized combat bomber which protected the B-29 crews from temperatures to 50° below zero. Its design helped protect the crew but engineers needed to solve the problem of opening the bomb-bay doors across the bottom of a pressurized aircraft. (U.S. Army Air Corps)

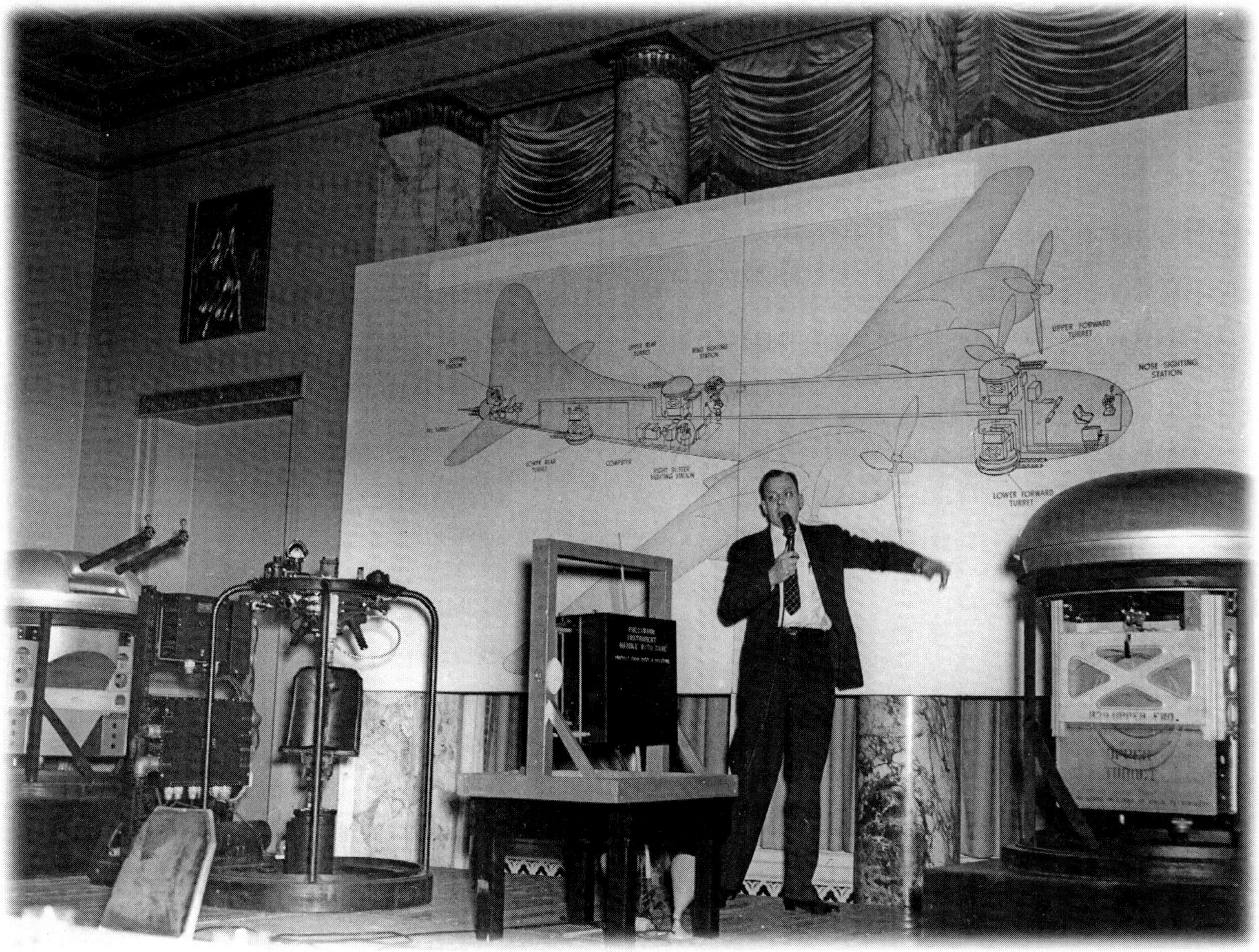

B-29 Superfortress – B.R. Prentice, a General Electric engineer, was photographed while he described the basic elements of the central B-29 gunfire control system to U.S. Army Officers. He explained, "All a gunner was supposed to do was get the enemy plane in his sight."[1] The complicated mathematical problem of accounting for speed, distance, gravity and parallax was worked out by the General Electric system. (General Electric) [1]General Electric Press Release that accompanied the photograph

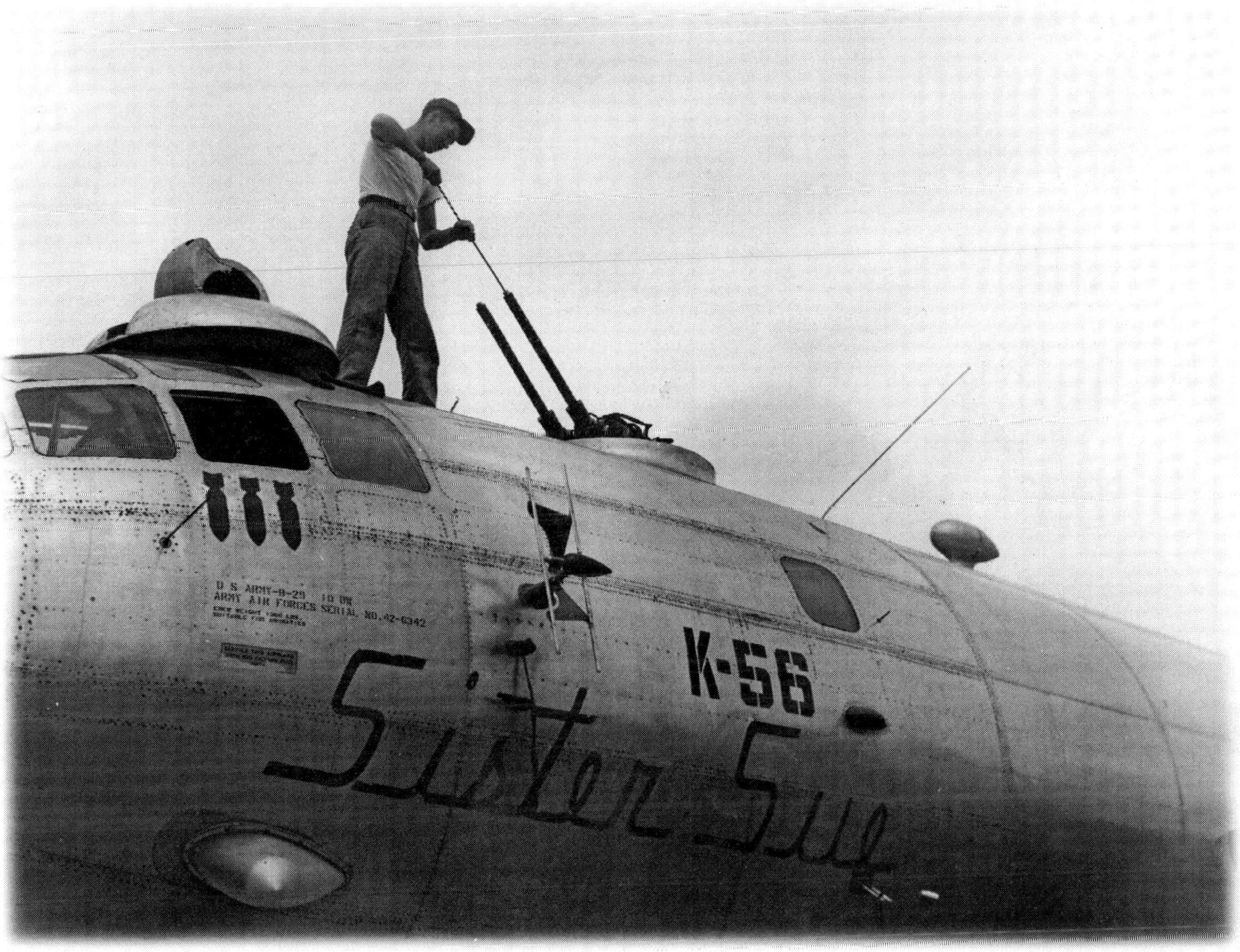

B-29 Superfortress – Sergeant Calvin Brown from Kimball, MN, was gunner aboard the B-29 *Sister Sue* and was photographed in January 1945 as he cleaned the upper forward turret guns. Sergeant Brown was well decorated with the Asiatic-Pacific Campaign Medal, American Theatre Service Medal, Good Conduct Medal, Air Offensive Japan, Burma-India-China Offensive, Air Combat India Burma, Presidential Unit Citation, and the WWII Victory Medal.[1] Sometime after this photograph was taken, the aircraft was lost in China; all but two of the crew escaped. (American Airlines) [1]58th Bomb Wing Memorial, New England Air Museum

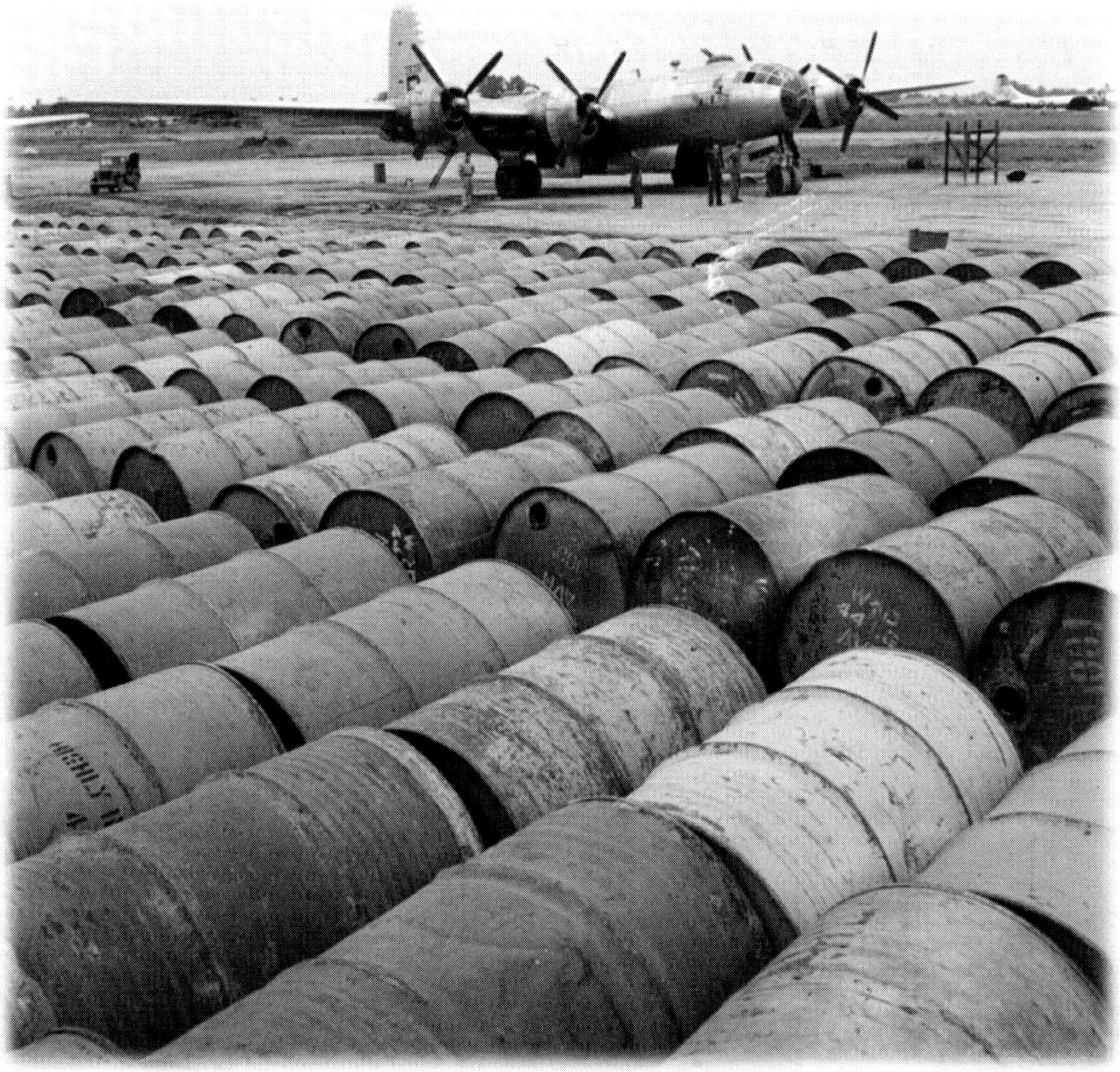

B-29 Superfortress – Somewhere in China, a B-29 bomber was being prepared for another raid. In the foreground are hundreds of gasoline drums that were used to refuel aircraft. Thousands of drums were needed and each was flown over the Hump of the Himalayas during the war. The B-29 was modified to hold 9,438 gallons of fuel and each drum held 55 gallons of gasoline which was usually pumped into the aircraft manually. (Bob Bryant)

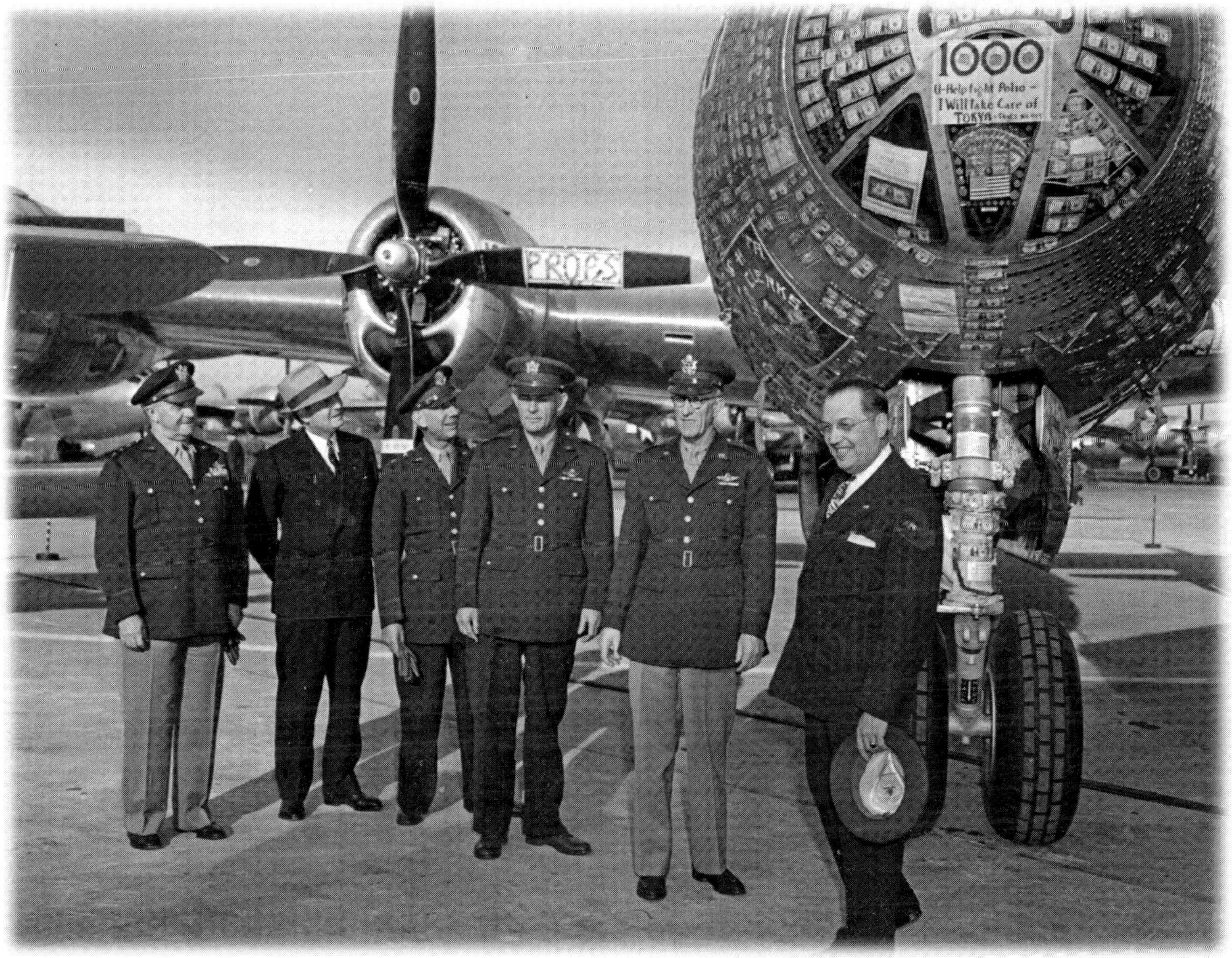

B-29 Superfortress – The highly decorated 1000th B-29 was delivered to the U.S. Army Air Force in Wichita, KS, in an impressive ceremony. Attending left to right: Lieutenant General Barton K. Yount, Commanding General, AAF Training Command; C.L. Egtvedt, Chairman and Chief Executive, Boeing Airplane Company; Major General K.B. Wolfe, Chief of Engineering and Procurement, ATSC; Major General Oliver P. Echols, Assistant Chief of Air Staff, Material and Services; Brigadier General Ray G. Harris, Commanding General, Midwestern District, ATSC and J.E. Schaefer, Vice-President and General Manager, Boeing, in Wichita. (Boeing Airplane Company)

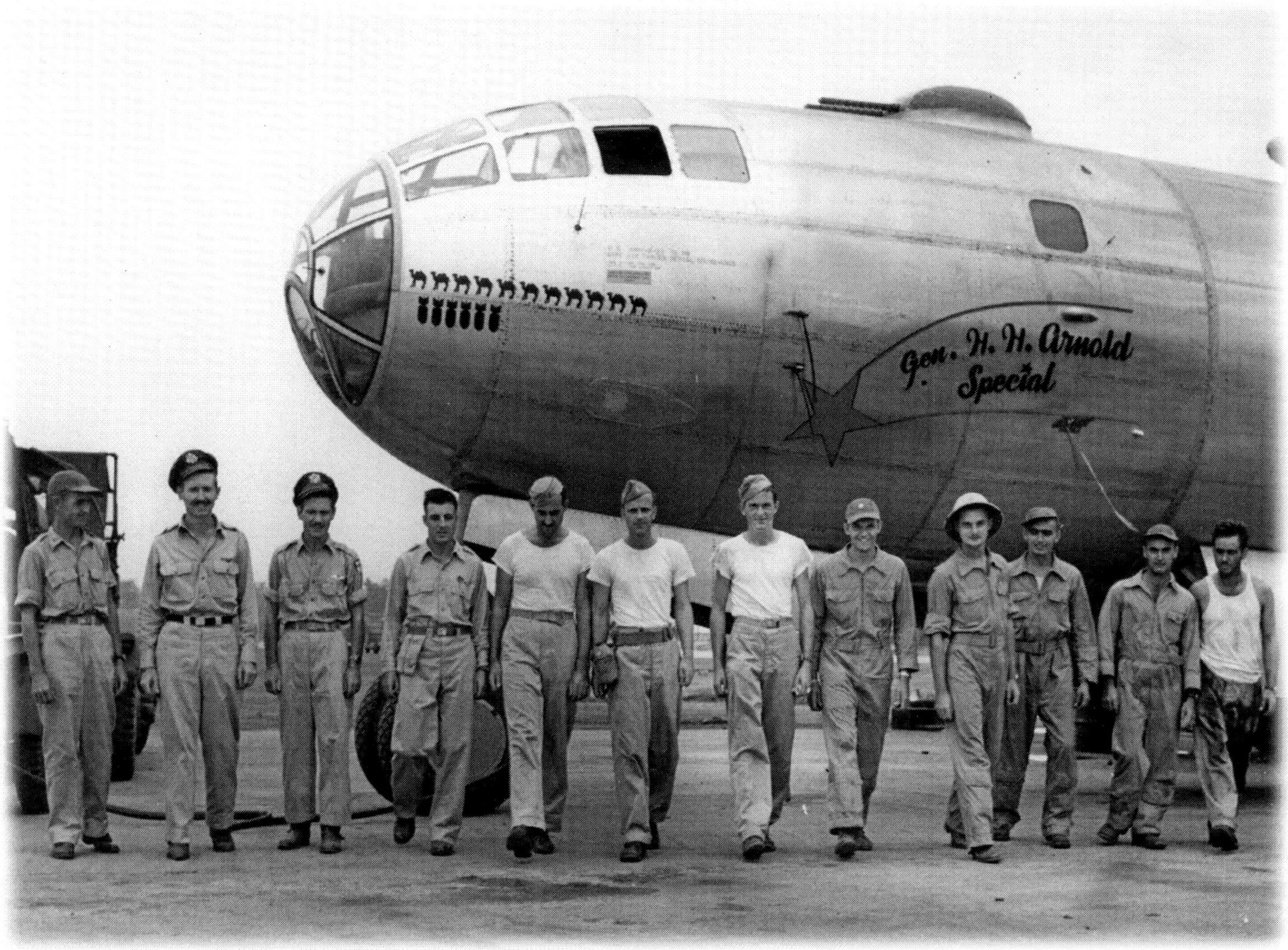

B-29 Superfortress – The aircraft pictured above was one of the initial 175 B-29s built. It was the 175th fuselage and was autographed by General Arnold at the Wichita Plant on January 11, 1944. The first 175 combat ready B-29s were no easy accomplishment either and became part of what was known as the Battle of Kansas. The bomber was destined for fame because it was one of the first to bomb Tokyo. Sometime after this photograph was taken, the bomber was one of the three B-29s that landed in Vladivostok, Russia due to emergencies. It was dismantled by the Russians, starting their successful ten year attempt to clone a B-29 escalating the Cold War. The Russian B-29 designation became known as the TU–4 and its NATO designation was BULL. (U.S. Army Press Censor 61208)

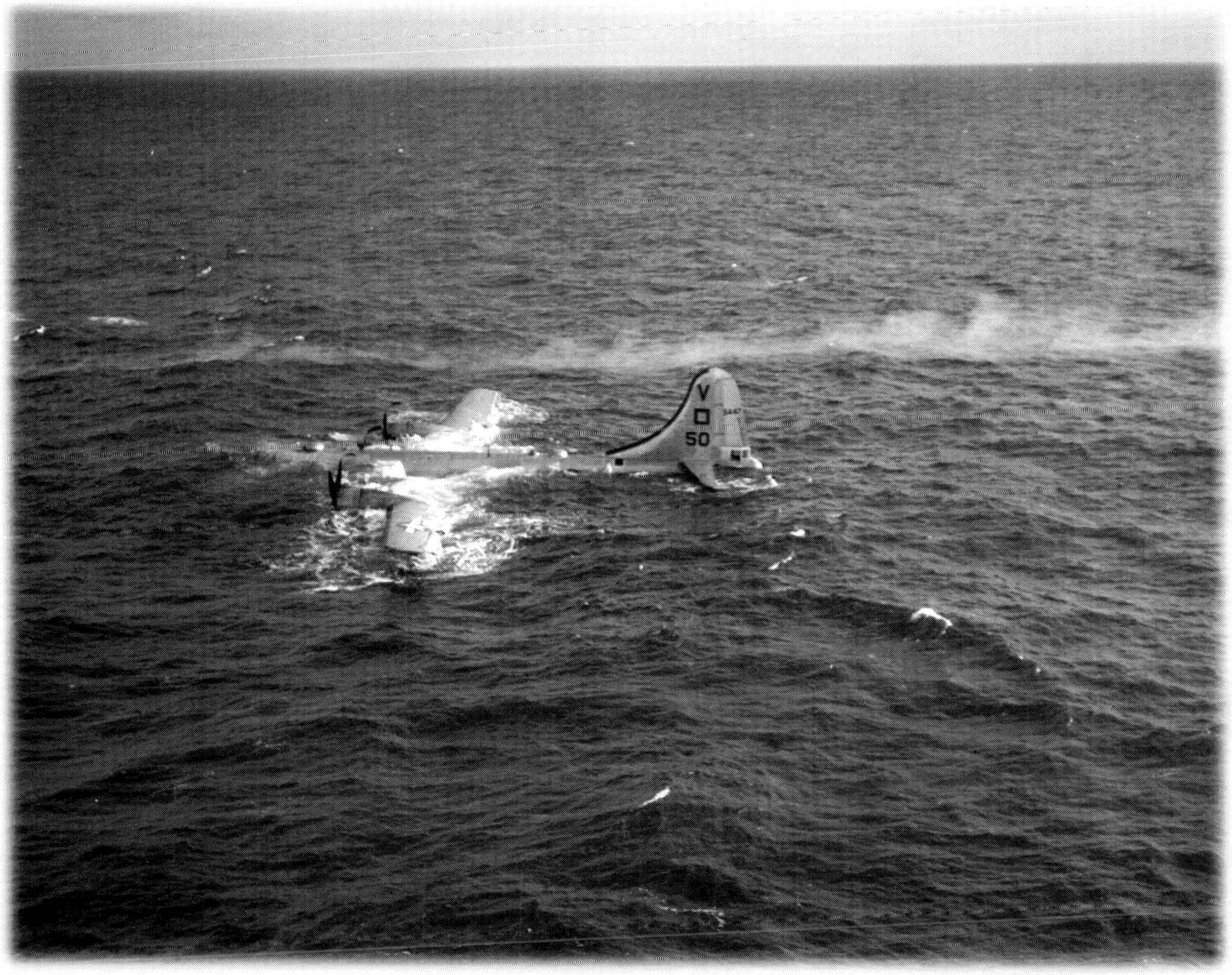

B-29 Superfortress – This B-29 raided Japan on Wednesday, December 13, 1944, and could not return to base. The crew was rescued by a U.S. Navy PBY Catalina. The distress signal smoke set by the crew was still visible as they evacuated the area. A sailor snapped a photograph from a rescue U.S. Navy PBY as the Pacific Ocean washed over the nose and much of the wing area of the big bomber. In a few seconds the entire aircraft would be submerged. (U.S. Navy)

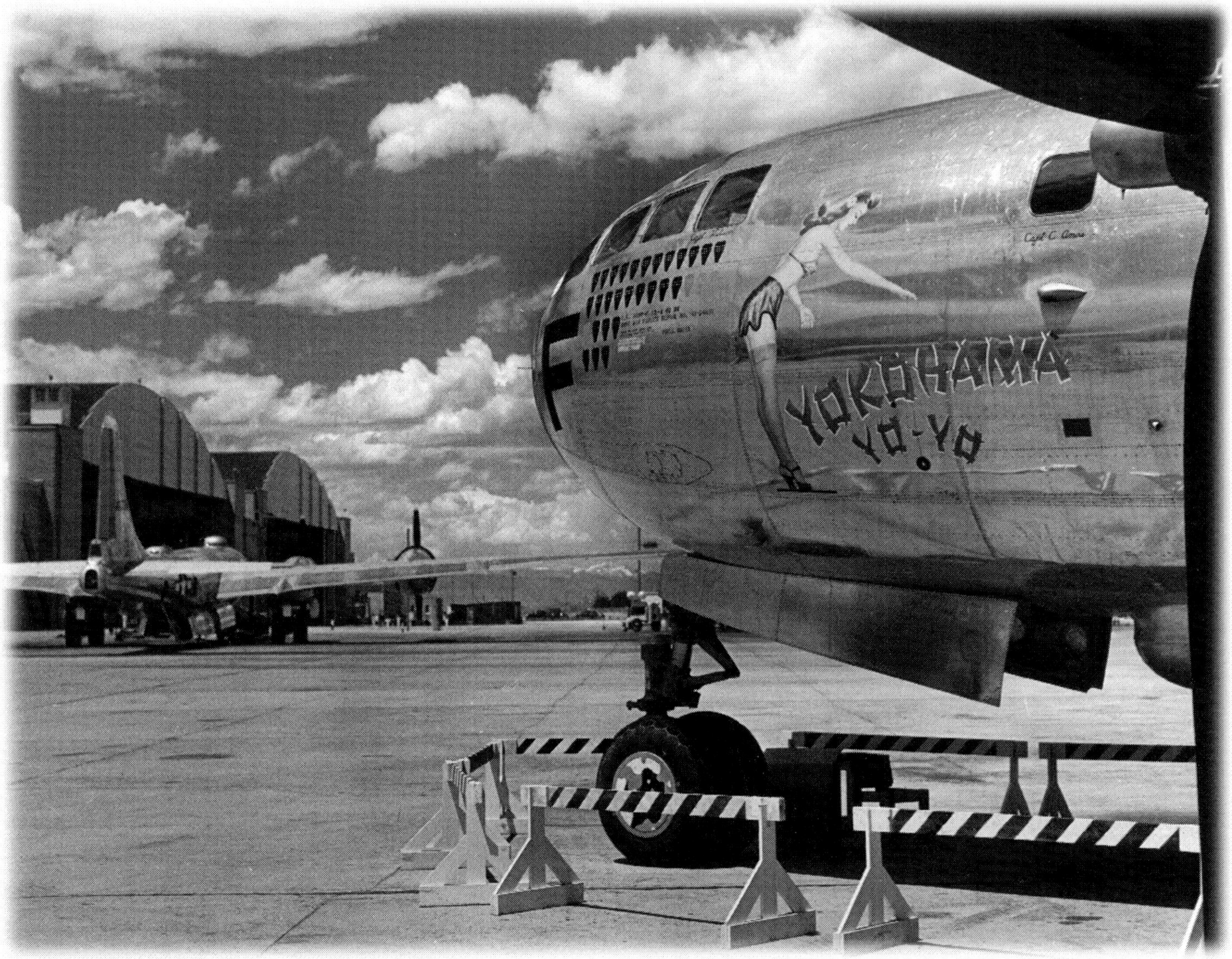

B-29 Superfortress – This B-29 was a heavily modified F-13-A photo reconnaissance aircraft. It was the second photo reconnaissance ship to fly over Tokyo. The aircraft had flown 30 missions including flying over Tokyo and Okinawa before their invasion by American forces. Its huge cameras, whose focal lengths reached 40 inches, clicked more than 30,000 exposures. The aircraft was photographed when it returned to the Denver Modification Center where technicians would study the aircraft fatigue after eight months of war-weary combat strain. The aircraft made the first non-stop flight from Honolulu to Denver, Colorado, in 14 hours and 3 minutes. (U.S. Army Air Corps)

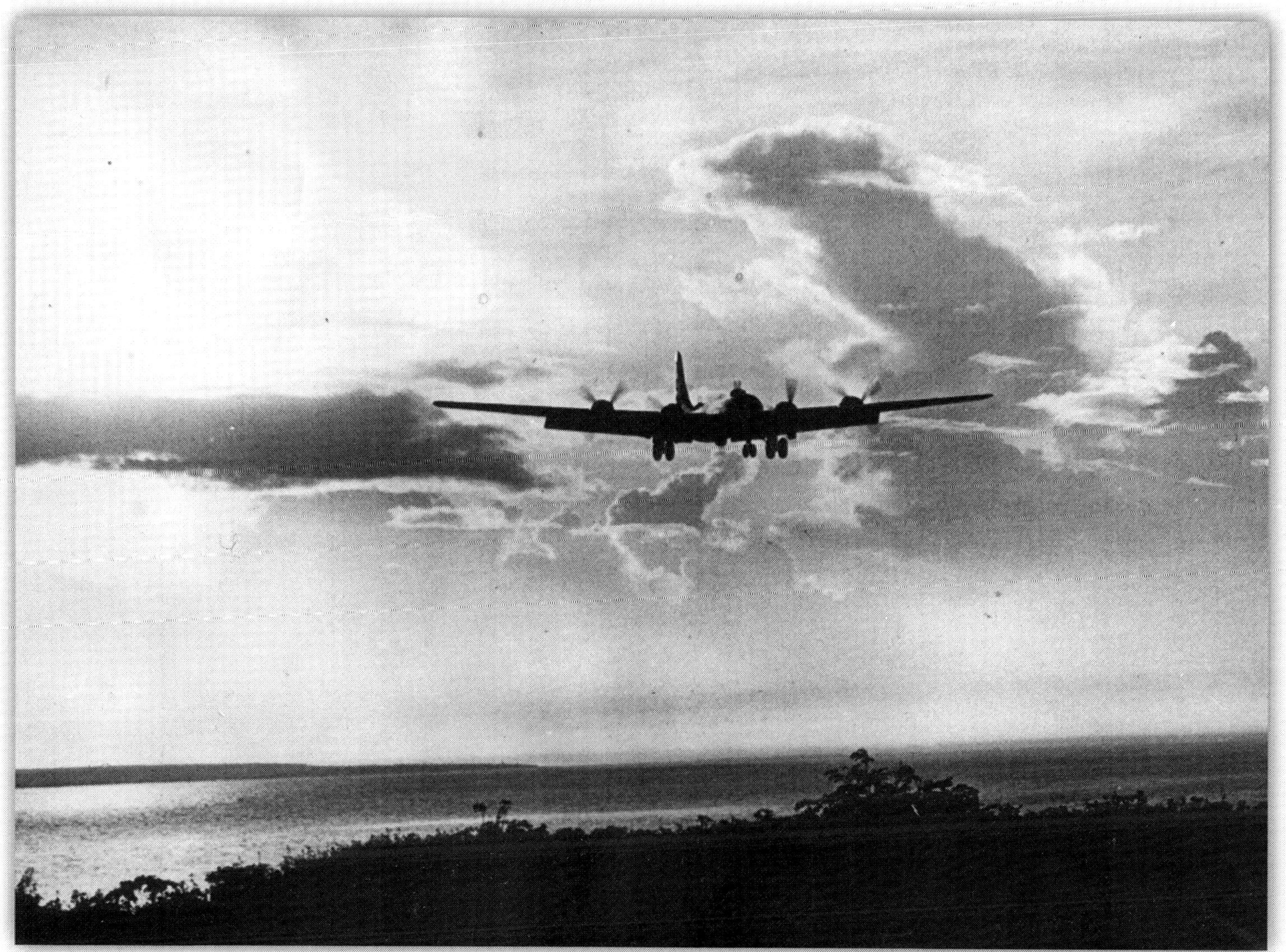

B-29 Superfortress – A B-29 from the 21st Bomber Command returns from a bombing raid on Tokyo. The aircraft was landing in Saipan against a western Pacific sunset. In July 1944, Curtis Le May was given command of the 21st Bomber Command which was responsible for the bombing of Japan. He believed that the country's entire manufacturing base, including the small cottage industries that fed the large factories of Japan's manufacturing industry, had to be destroyed if America was to launch a successful invasion of Japan.[1] Due to B-29 bombing missions, including dropping the world's first two nuclear bombs, no invasion of the Japanese home islands was necessary. Japan formally surrendered on September 2, 1945. (U.S. Army Air Corps) [1]www.historylearningsite.co.uk

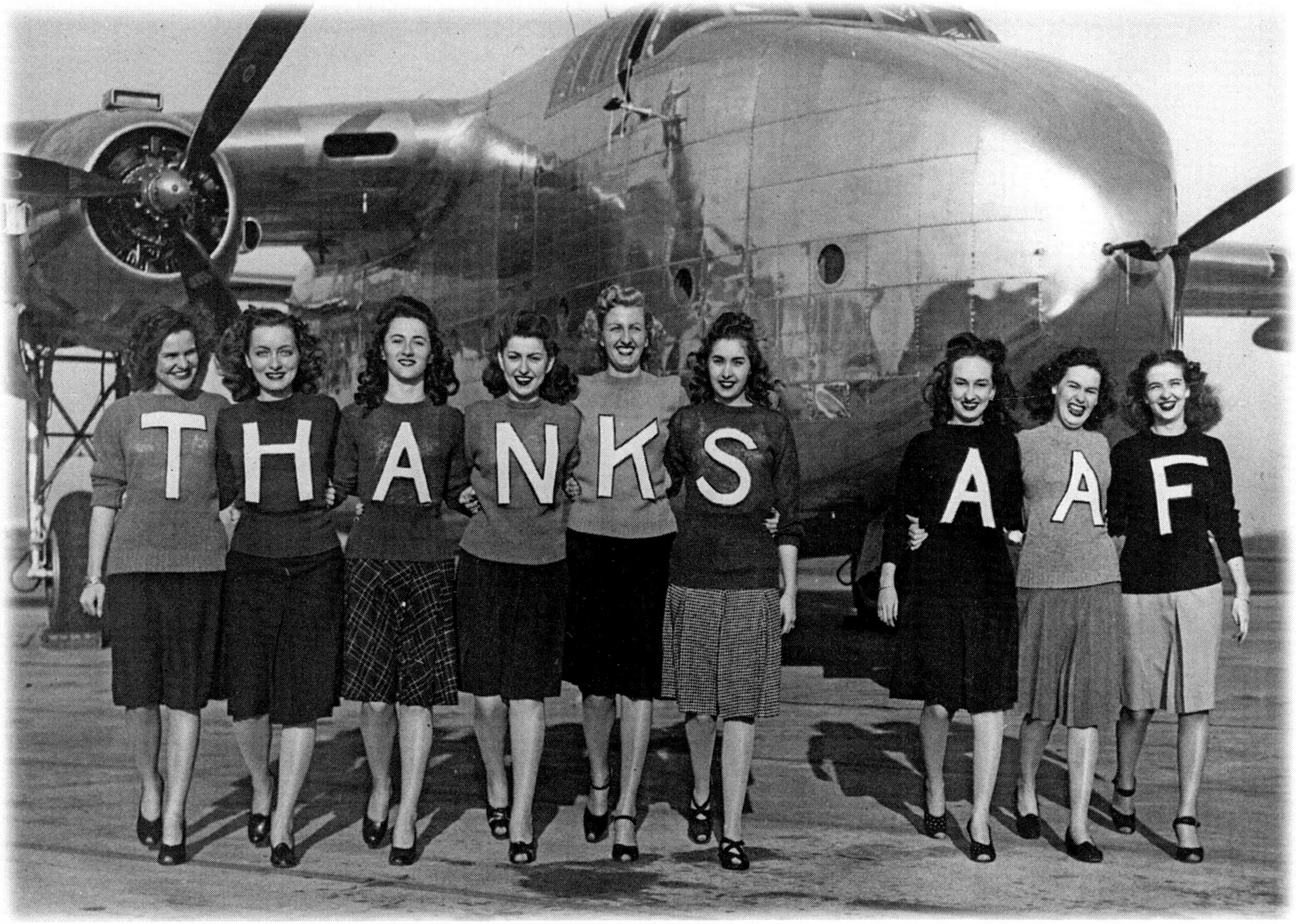

WWII is Over – Staff workers from the Washington War Finance Committee spelled out their sentiments at the Wings of Victory exhibition event, thanking the Army Air Forces for their service in WWII. The exhibition was held at the Washington National Airport one month after the Japanese formally surrendered bringing WWII to a close. The War Finance Committee and War Advertising Council had been responsible for promoting voluntary compliance with bond buying quotas and rationing programs. (U.S. Army Air Forces)

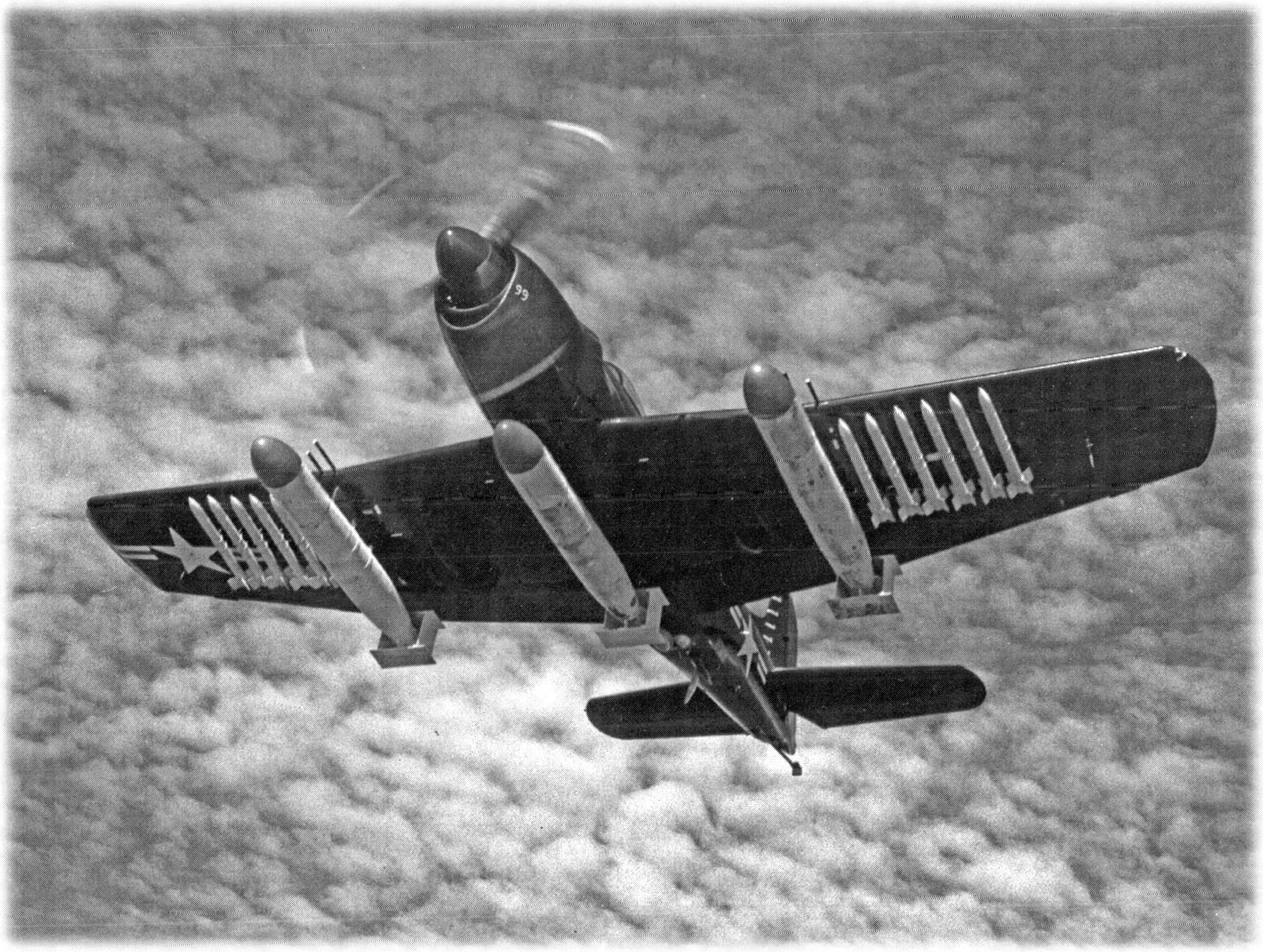

AM-1 Mauler – The Martin AM-1 Mauler fighter bomber was too late for WWII and ended production before the Korean War. Although powerful, first production AM-1 Maulers did not fly until December 1946 and its deliveries to Navy squadrons didn't start until March 1948. Photographed in January 1949, this Mauler could carry a payload of 9,000 pounds. A B-17 carried 8,000 pounds on short range missions. Only 149 Maulers were completed before production ceased in October 1949 [1] (U.S. Navy) [1]www.daveswarbirds.com

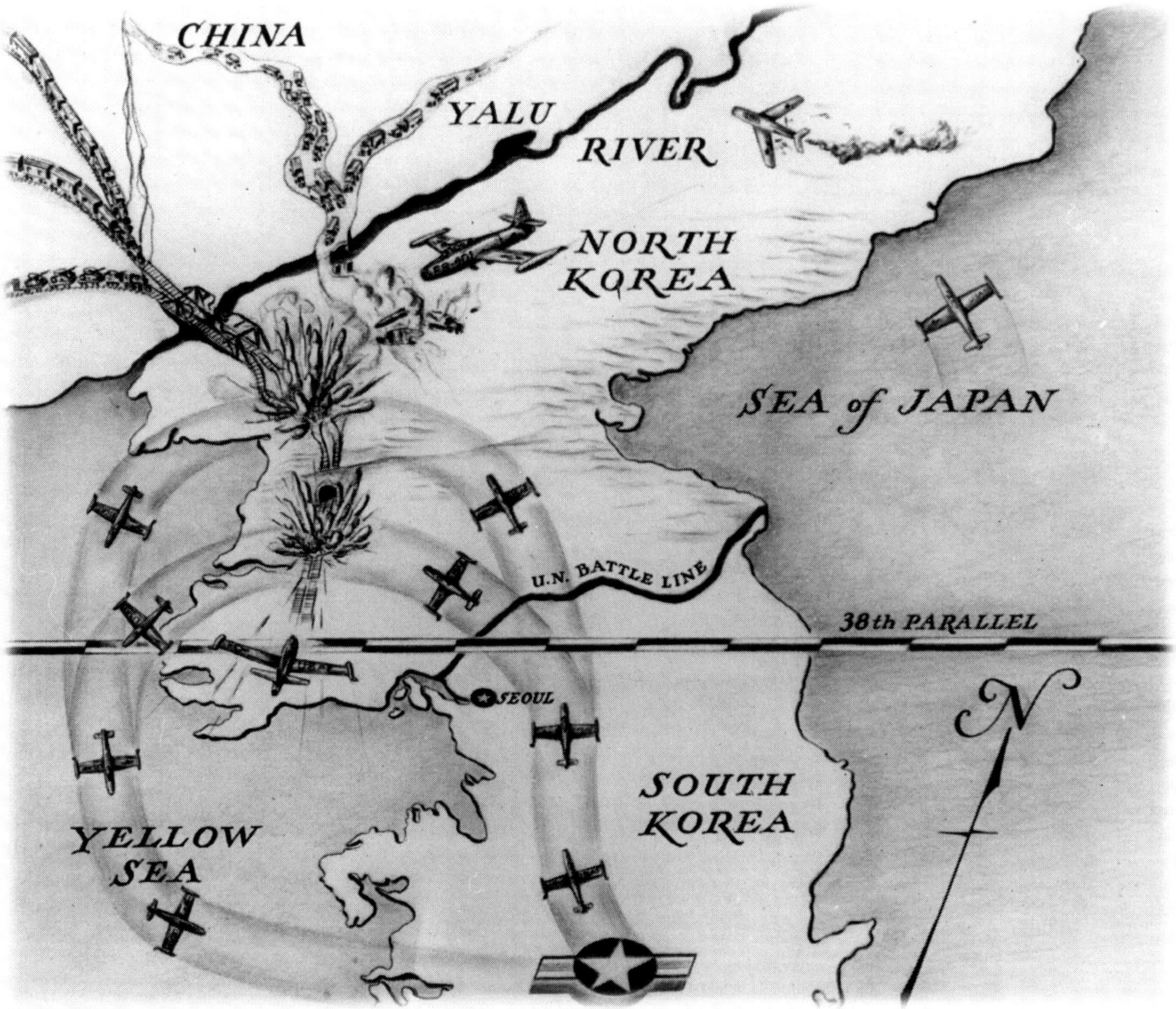

Korean War – The Korean War was fought between 1950 and 1953 and represented combat between an American-led United Nations force, North Korea and China. It officially ended when President Dwight D. Eisenhower signed an armistice on July 27, 1953. When the war ended, boundaries of North and South Korea were the same as before the costly war. Over 54,000 Americans and well over one million Korean and Chinese died in the conflict. (Republic Aviation Corporation)

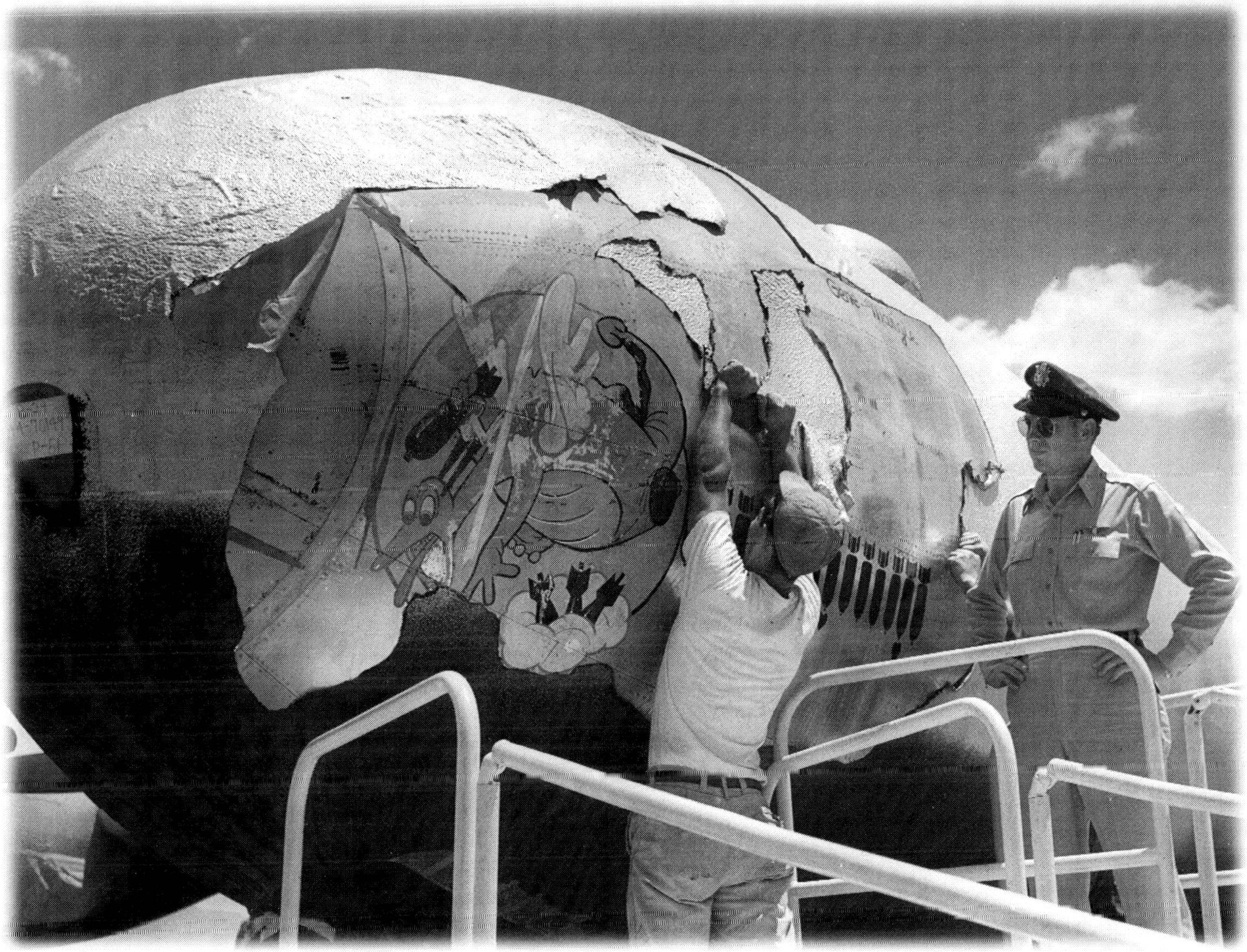

B-29 Superfortress Returns to the Air – Thousands of B-29s survived WWII only to be moth-balled in the American desert at the Davis-Monthan Air Base. As the Korean Conflict escalated, it was apparent that heavy bombers would once again be needed. The B-29 was selected to return to the air instead of the new B-36 bomber. B-29s were recovered from their desert sleep and put back into service. Captain Howard L. Hensley was the Public Information Officer at Davis-Monthan when this B-29 had its protective covering removed. (Downie)

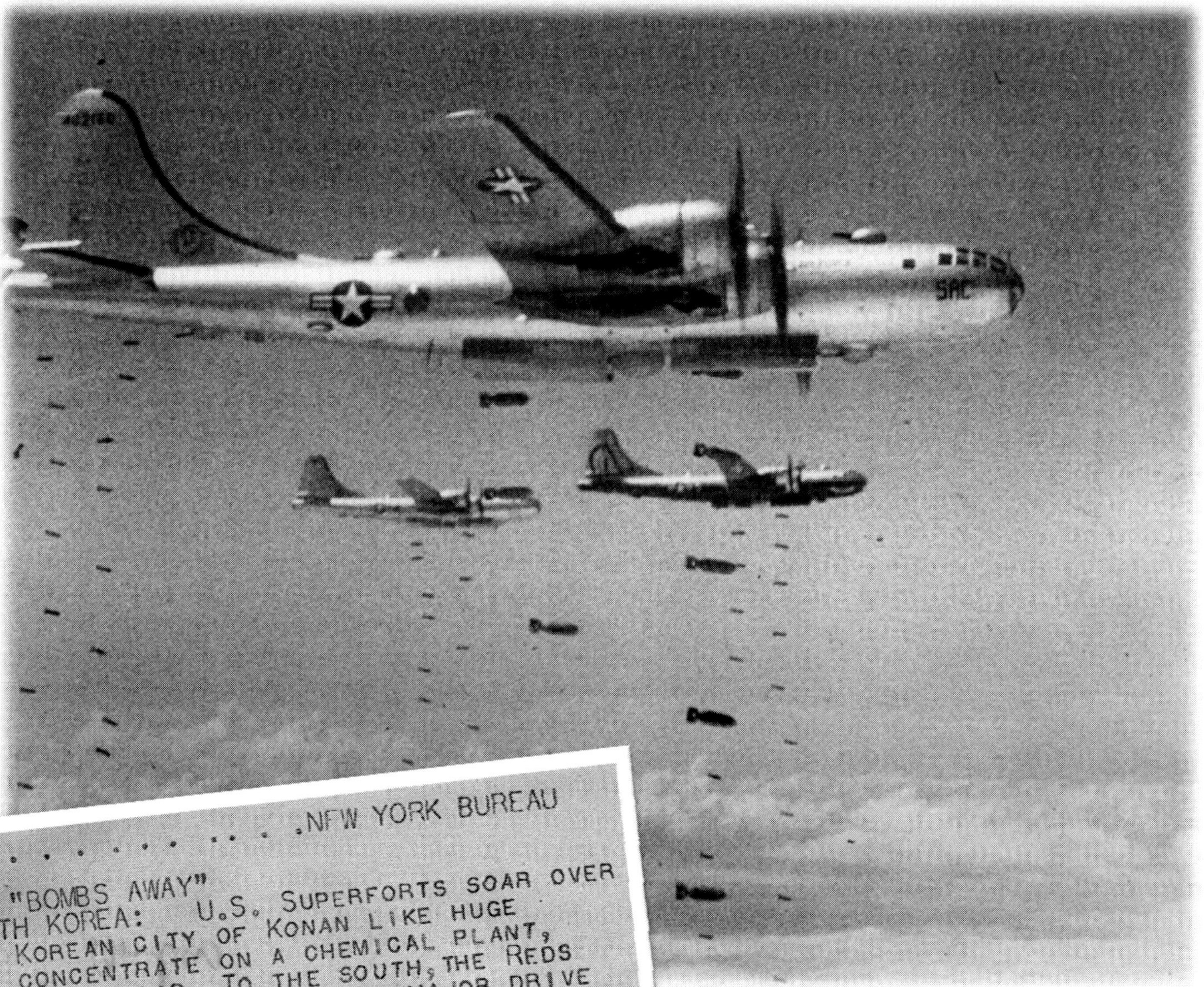

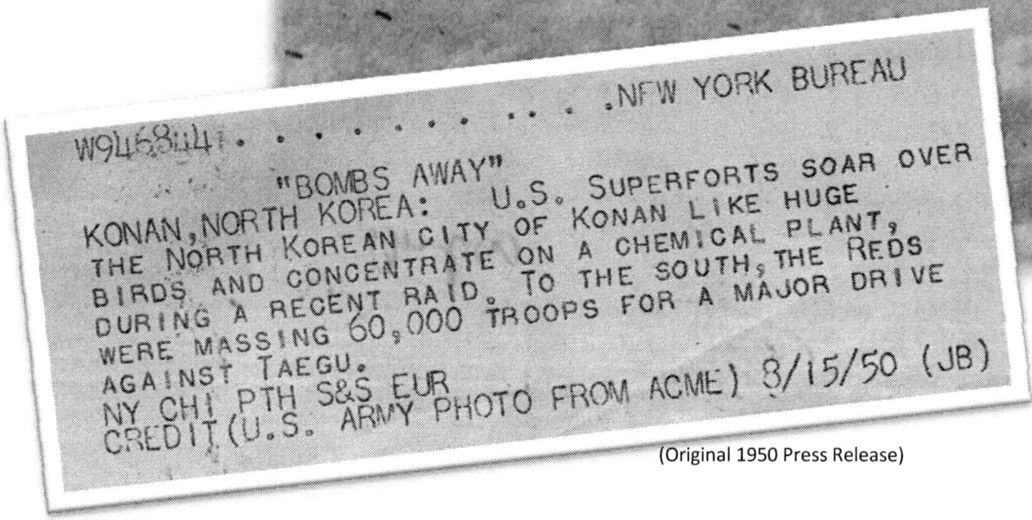

W9468441 . . . . NEW YORK BUREAU

"BOMBS AWAY"

KONAN, NORTH KOREA: U.S. SUPERFORTS SOAR OVER THE NORTH KOREAN CITY OF KONAN LIKE HUGE BIRDS AND CONCENTRATE ON A CHEMICAL PLANT, DURING A RECENT RAID. TO THE SOUTH, THE REDS WERE MASSING 60,000 TROOPS FOR A MAJOR DRIVE AGAINST TAEGU.

NY CHI PTH S&S EUR

CREDIT (U.S. ARMY PHOTO FROM ACME) 8/15/50 (JB)

(Original 1950 Press Release)

B-29 Superfortress – Similar to scenes from WWII, hundreds of B-29s took to the air in the Korean War. Rather than face aged Zero Japanese fighters, the crews of B-29s now faced Russian MiG-15 jet fighters who could reach their altitude. (U.S. Army & ACME)

OFF ON MIDNIGHT MISSION

CAPT. WILLIAM K. HOOK (IN COCKPIT), OF NORTH ST. PETERSBURG, FLA., REVS UP THE ENGINE OF HIS F-51 MUSTANG FIGHTER PLANE BEFORE TAKEOFF FROM A U.S. FIFTH AIR FORCE ADVANCED BASE IN KOREA ON A NIGHT MISSION AGAINST A NORTH KOREAN CONVOY. PLANE IS ARMED WITH ROCKETS, BOMBS AND MACHINE GUNS. MAN AT RIGHT IN COCKPIT IS NOT IDENTIFIED.

(Original Press Release)

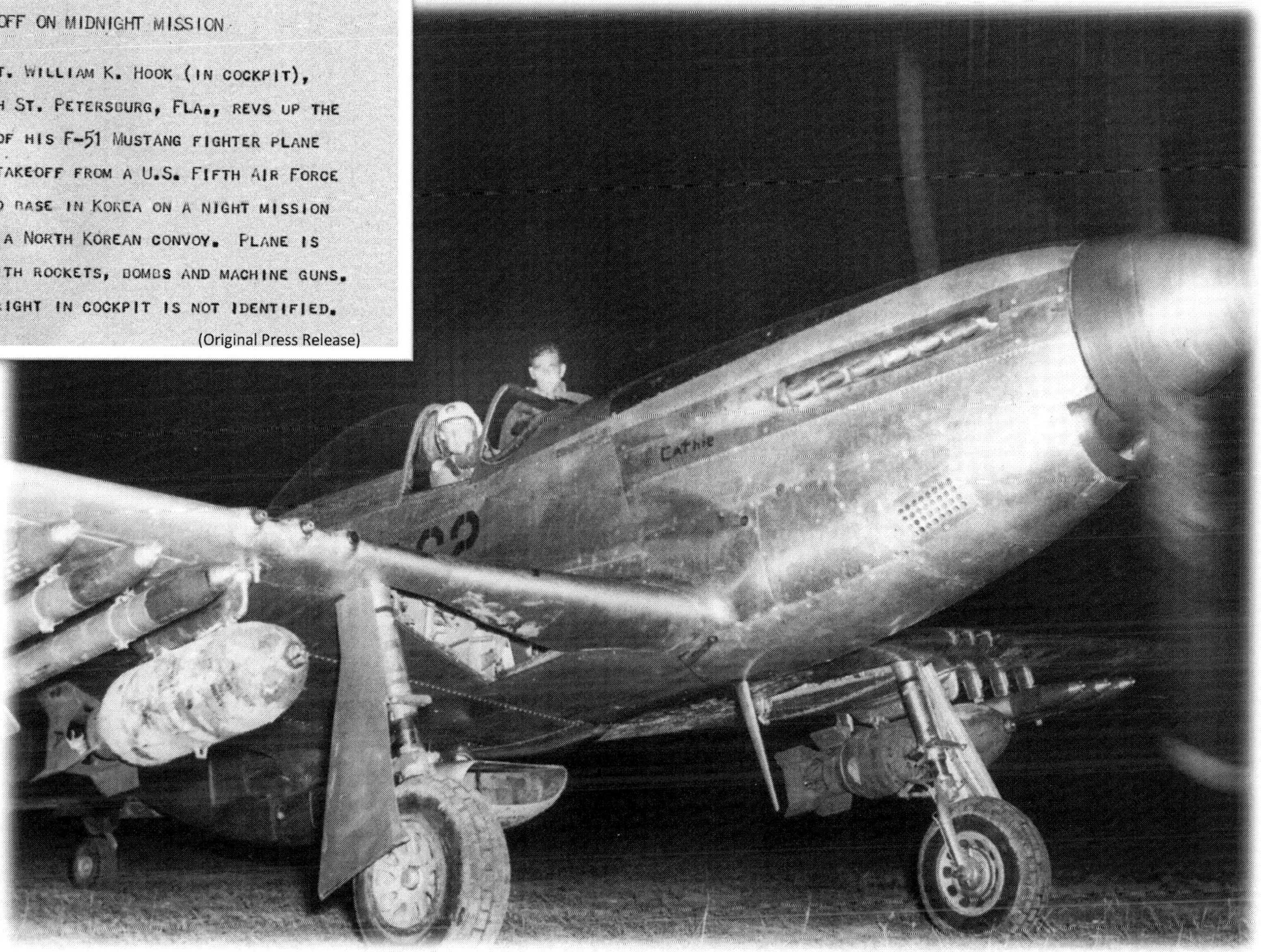

P-51 Mustang – Many WWII aircraft were once again put into service during the Korean War. The P-51 Mustang was not a bomber but rather a fighter that could carry bombs. As in WWII, the P-51 Mustang proved itself a versatile aircraft especially compared to early jet fighter bombers. The early jet aircraft had limited range while the Mustang added valuable staying power to the aerial NATO armada. (U.S. Air Force Strategic Air Command)

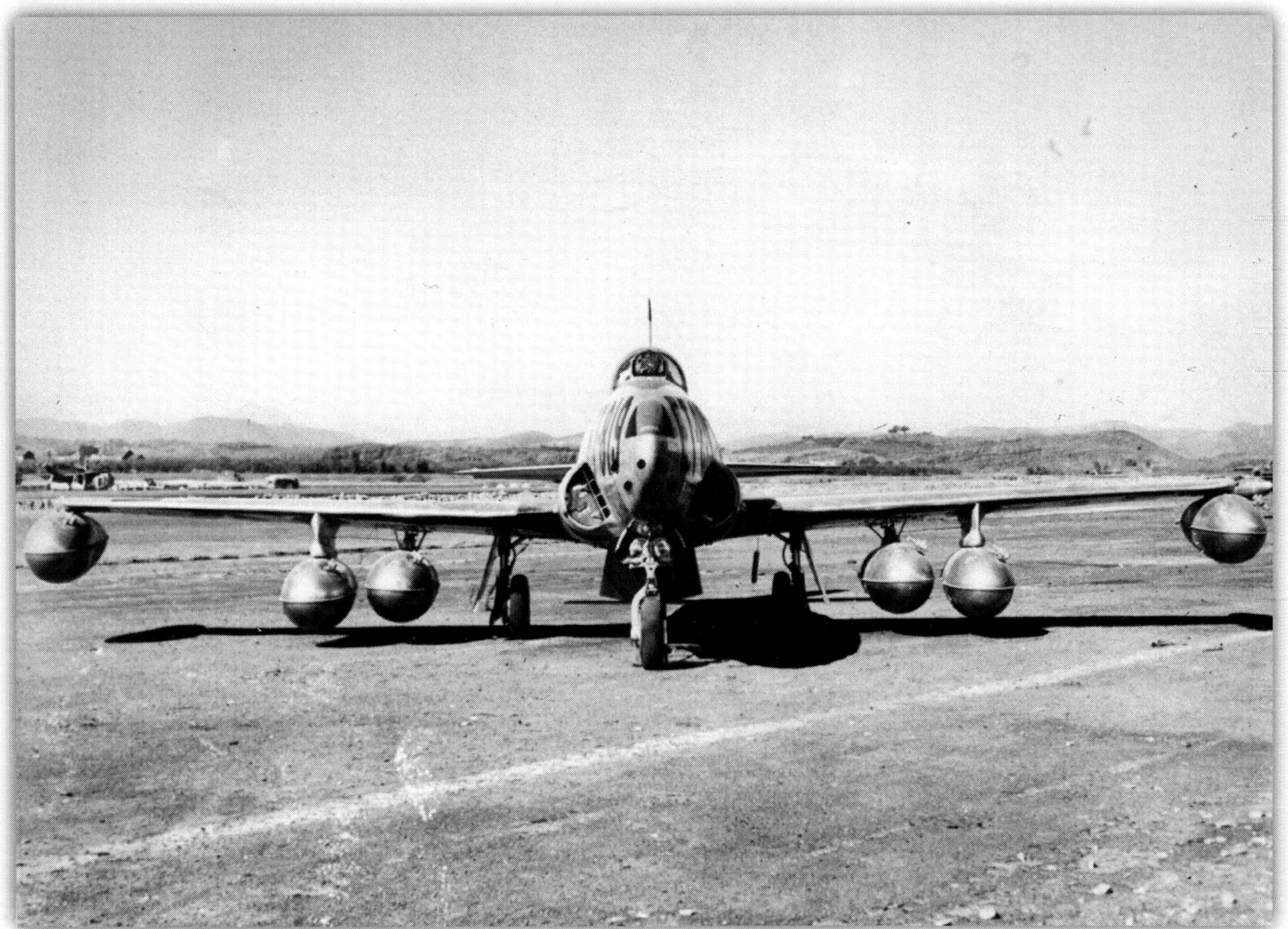

F-80 – The Lockheed F-80 first flew in 1944 and was one of the world's first successful turbojet-powered combat aircraft. It proved to be an able fighter bomber in Korea. Shown in this February 1951 photograph, the F-80 could carry six deadly 110 gallon Napalm tanks. The bomb struck and skipped as it engulfed the target in flame. (International News Photograph)

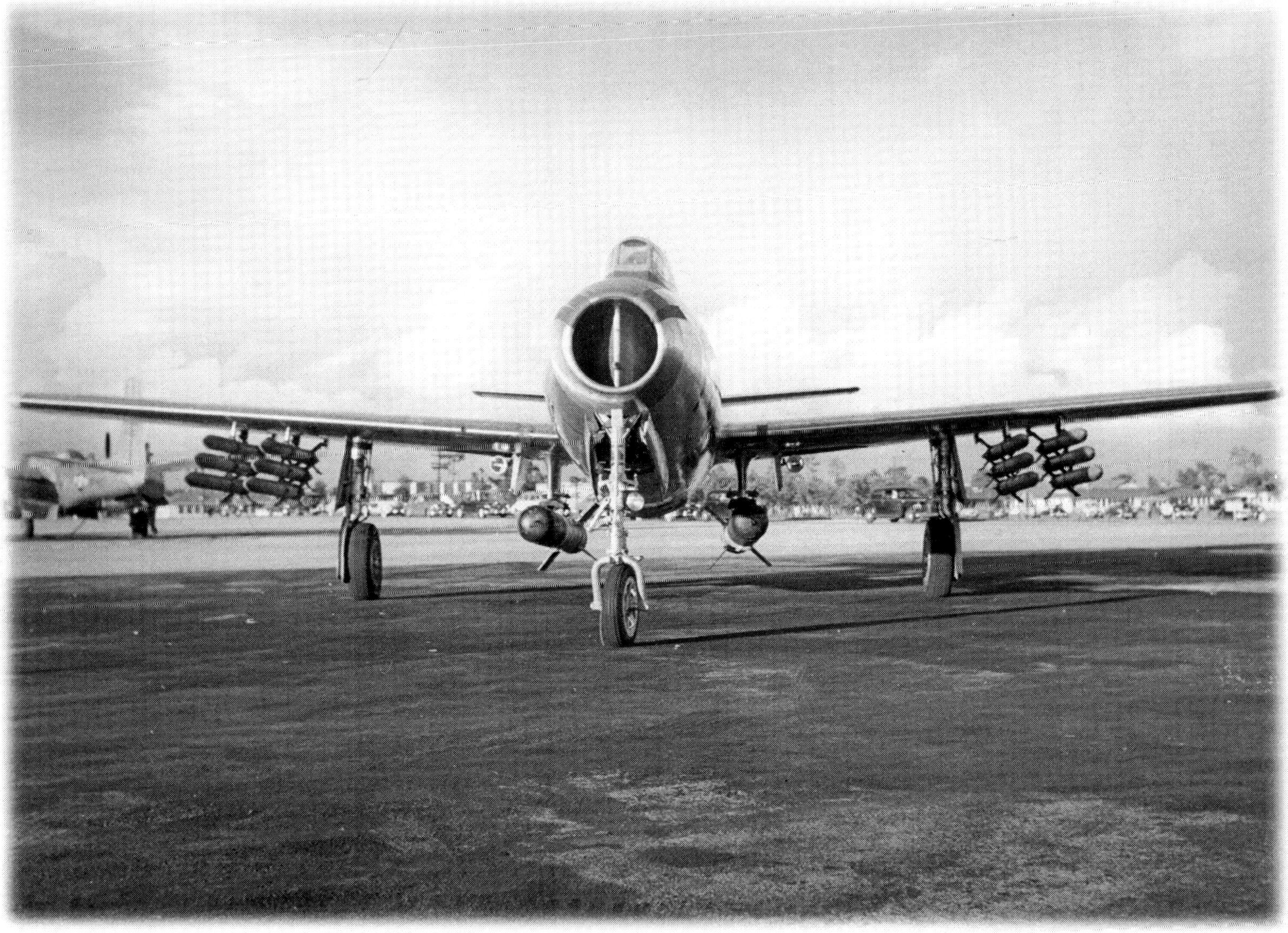

F-84 Thunderjet – Republic's F-84B Thunderjet joined In the nation's defense in December 1947 and was quickly upgraded to the D model for structural issues. F-84D models initially protected Boeing B-29 Superfortresses against North Korean fighters. As other jet fighters became available, it was moved to a successful ground assault role. The F-84 was responsible for dropping 50,427 tons of ordnance on enemy positions during the Korean War.[1] The F-86E pictured above featured an elongated cockpit and advanced radar-ranging gunsight. The pictured E model became available after the Korean War and helped keep America safe during the early years of the Cold War. (Republic Aviation) [1]*Republic F-84 Thunderjet / Thunderstreak / Thunderflash Fighter Bomber* By Dan Alex 2010

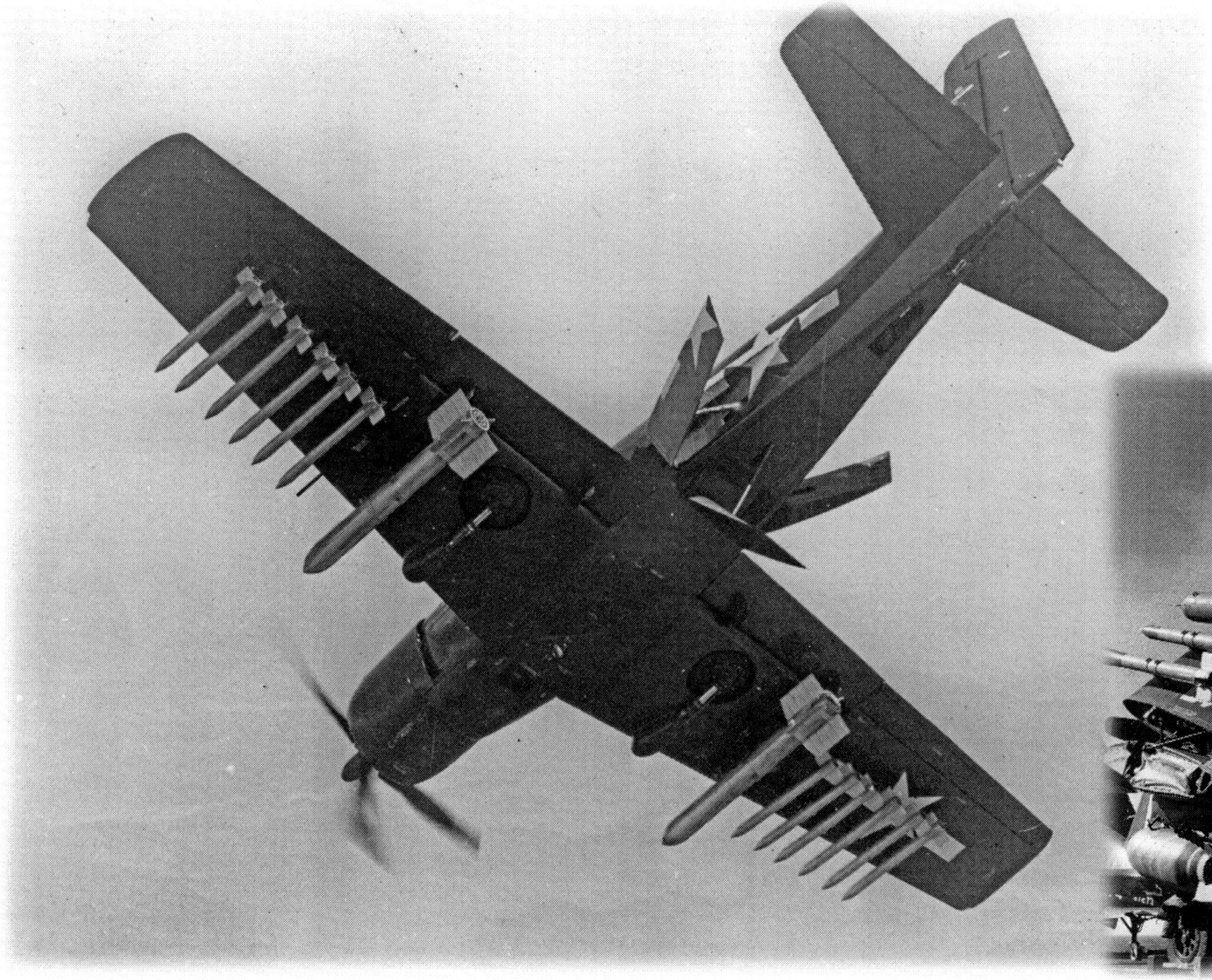

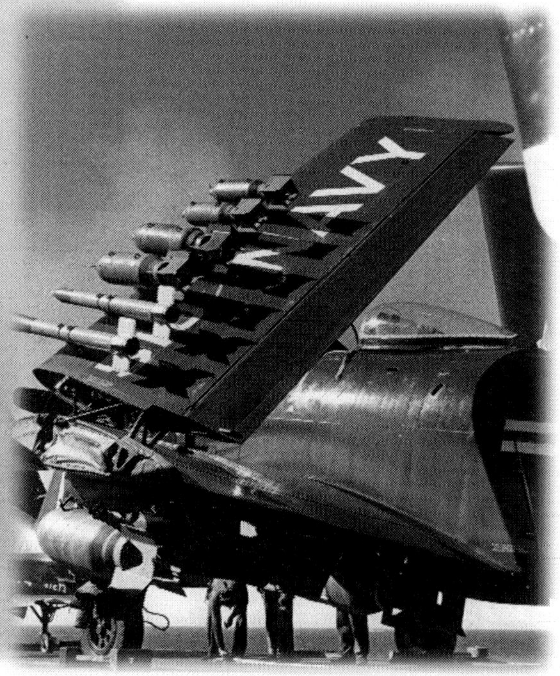

AD Skyraider – The AD Skyraider was designed during WWII as a follow-on replacement to earlier Navy Helldiver and Avenger dive and torpedo bombers. The aircraft joined the Korean War from the deck of the *USS Valley Forge* on July 3, 1950.[1] The aircraft hosted special fuselage flaps which opened from the sides and bottom of the aircraft and acted as brakes in combat maneuvering. The dive bomber was powerful and could carry 8,000 pounds of weapons and fly ten hours, exceeding contemporary jets of the time. Seven versions of the aircraft went into production for the Navy and Marines. The inset photograph is an AD-4 awaiting its next mission. (Douglas Aircraft Company, Inc.) [1]Mersky, Peter B. *U.S. Marine Corps Aviation: 1912 to the Present*. Annapolis, Maryland: The Nautical and Aviation Publishing Company of America, 1983. ISBN 0-933852-39-8

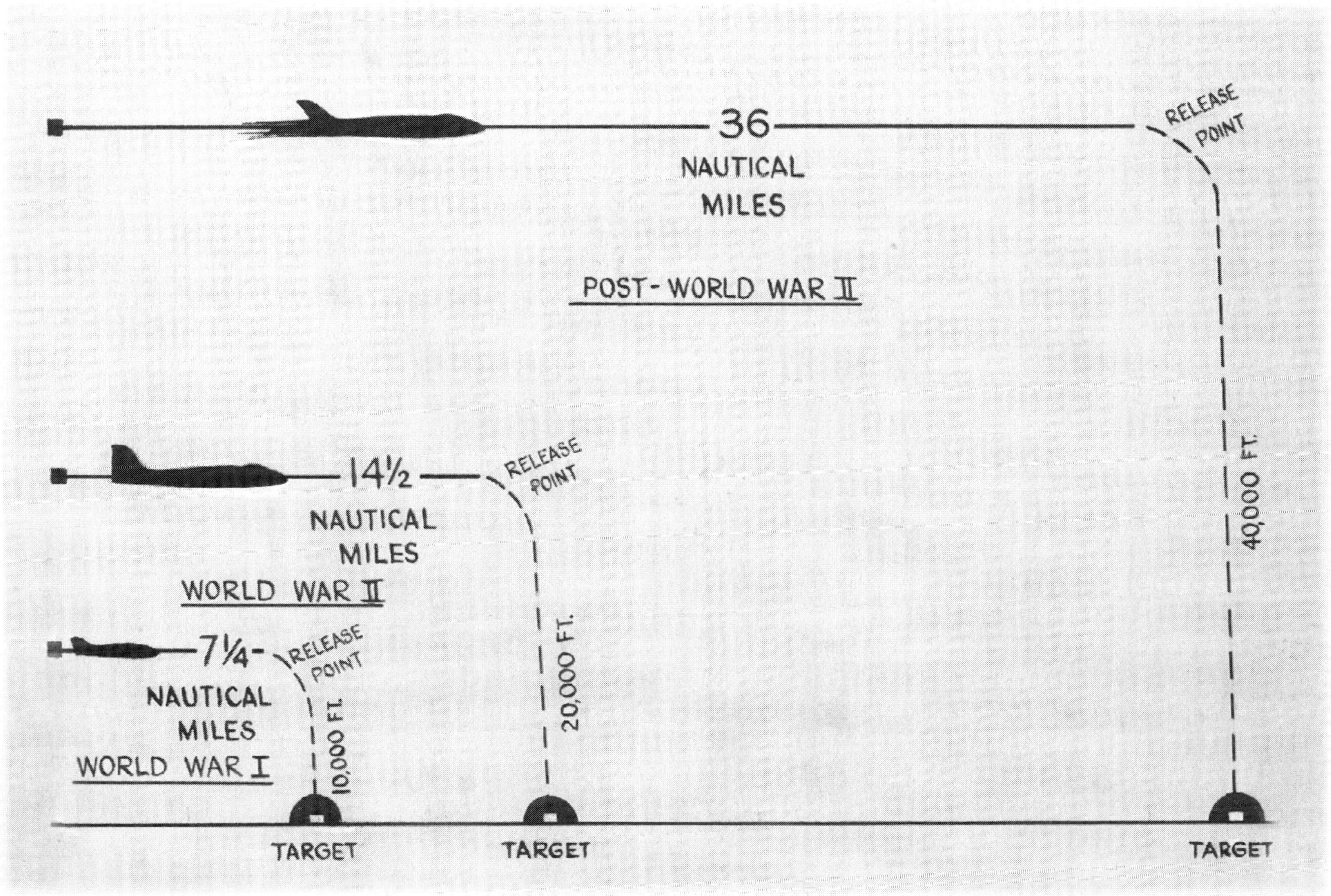

The Science of Bombing – This post-WWII illustration shows some of the decisive improvements made by bombers: higher altitudes, further distance from the target and faster speeds of aircraft as they delivered their payloads. Because of these changes bombers continued being important as the Cold War escalated. The technology of dropping conventional gravity directed explosives from aircraft continued its evolution but the advent of intercontinental missiles was a growing concern. (*Skyways Magazine* undated illustration drawing)

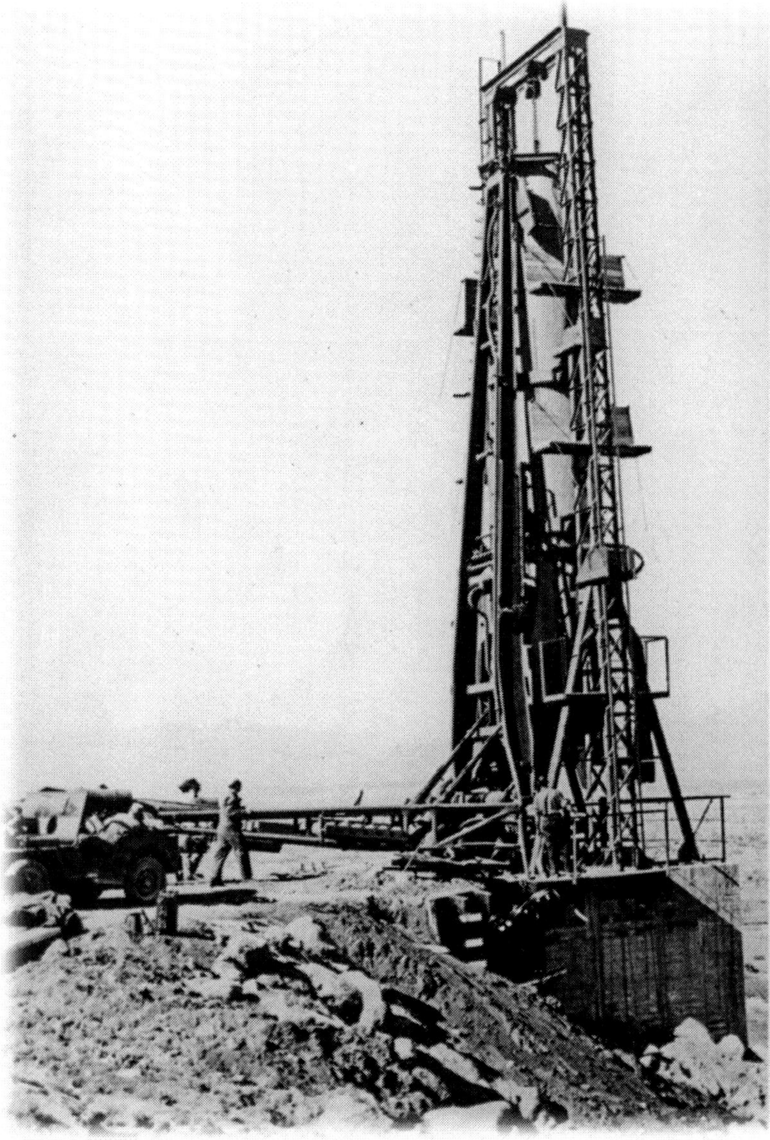

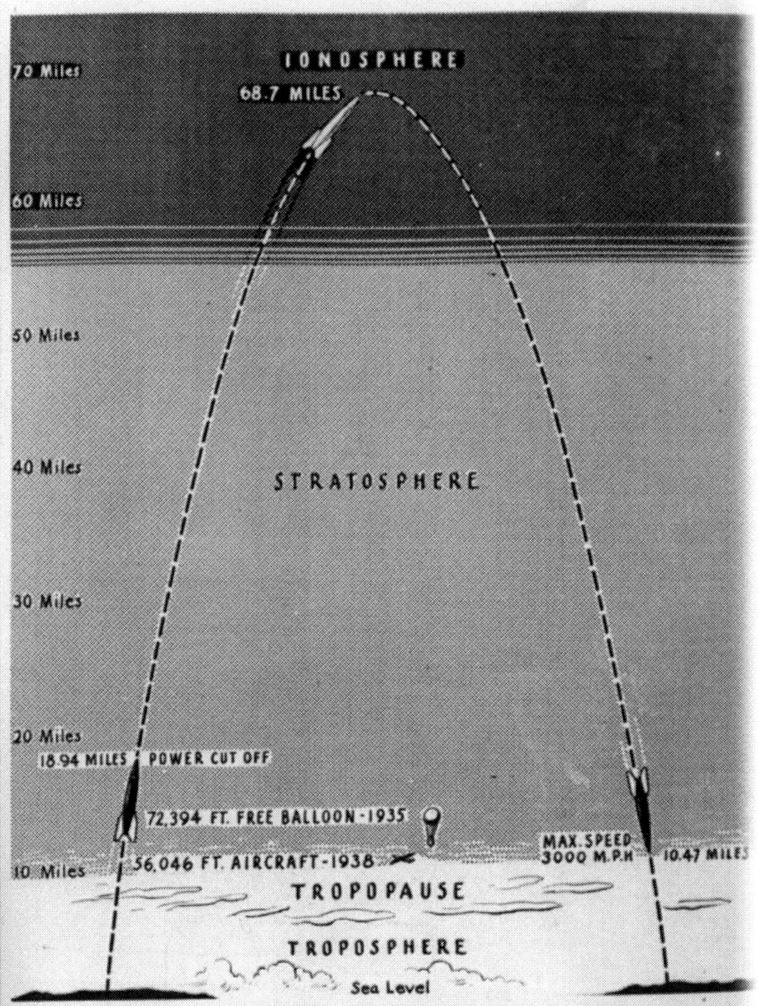

The Cold War – Most historians agree that the Cold War began before the Korean War. As the photograph above shows, NATO attempted to leverage their captured WWII German rocket and jet technologies. The Russians also leveraged their captured knowledge but no one knew who had captured the most important secrets. These images depict a U.S. launch projection from May 1946 that used a "mechanical brain." The 1946 computer required a ten day processing window to calculate the trajectory, which would have taken mathematicians three months. This V2 rocket launch carried scientific equipment instead of a warhead. (Both photographs by International News)

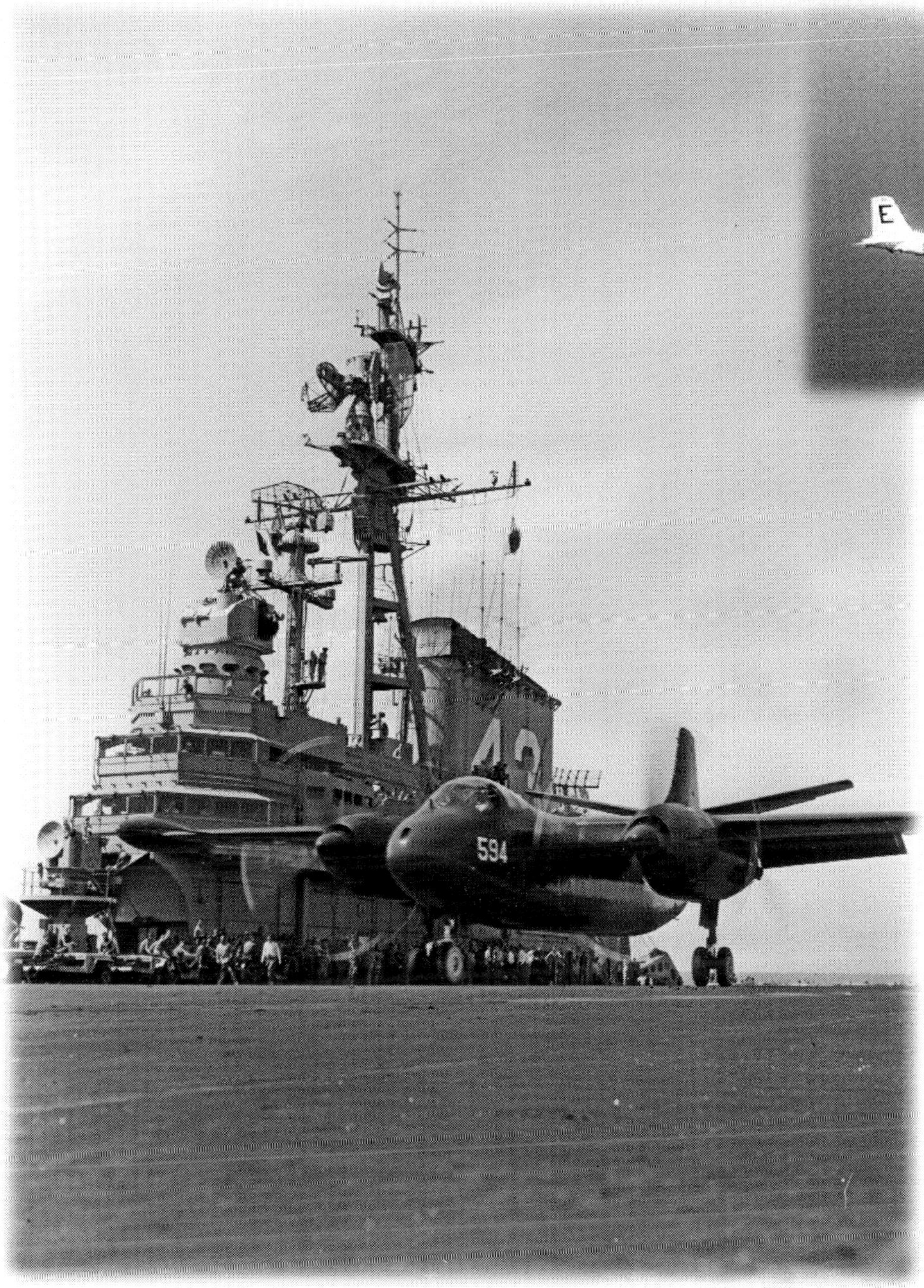

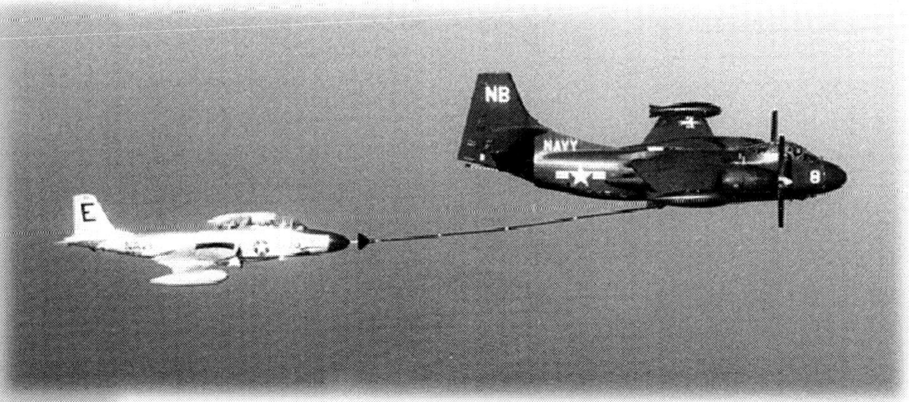

AJ-1 & 2 Savage

North American Aviation built the 27,558 pound nuclear bomb carrying AJ-1 aircraft. Fully loaded, the aircraft could weigh in at 51,000 pounds. In an age of jets, the AJ aircraft served in the U.S. Navy from 1948 until 1956. With a crew of three, carrying 12,000 pounds of conventional bombs or one nuclear bomb and with maximum speeds of 470 mph, it was a formidable carrier bomber. The early models AJ-1s, like the aircraft left, were converted into fuel tankers during the early 1950s. An AJ-2 is pictured below; look closely and you can see one of the crew standing out of the cockpit during taxi. (All photographs by U.S. Navy)

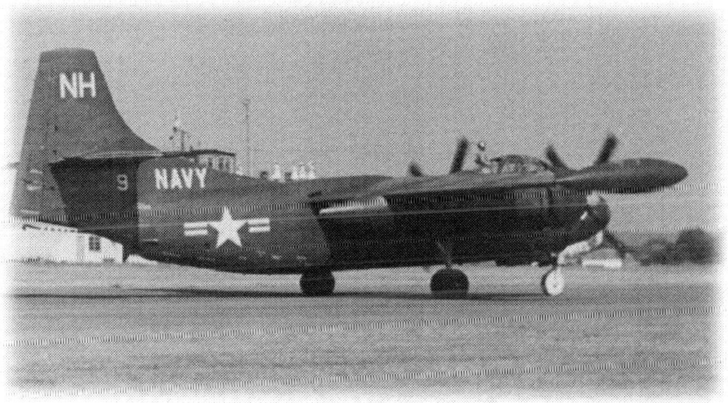

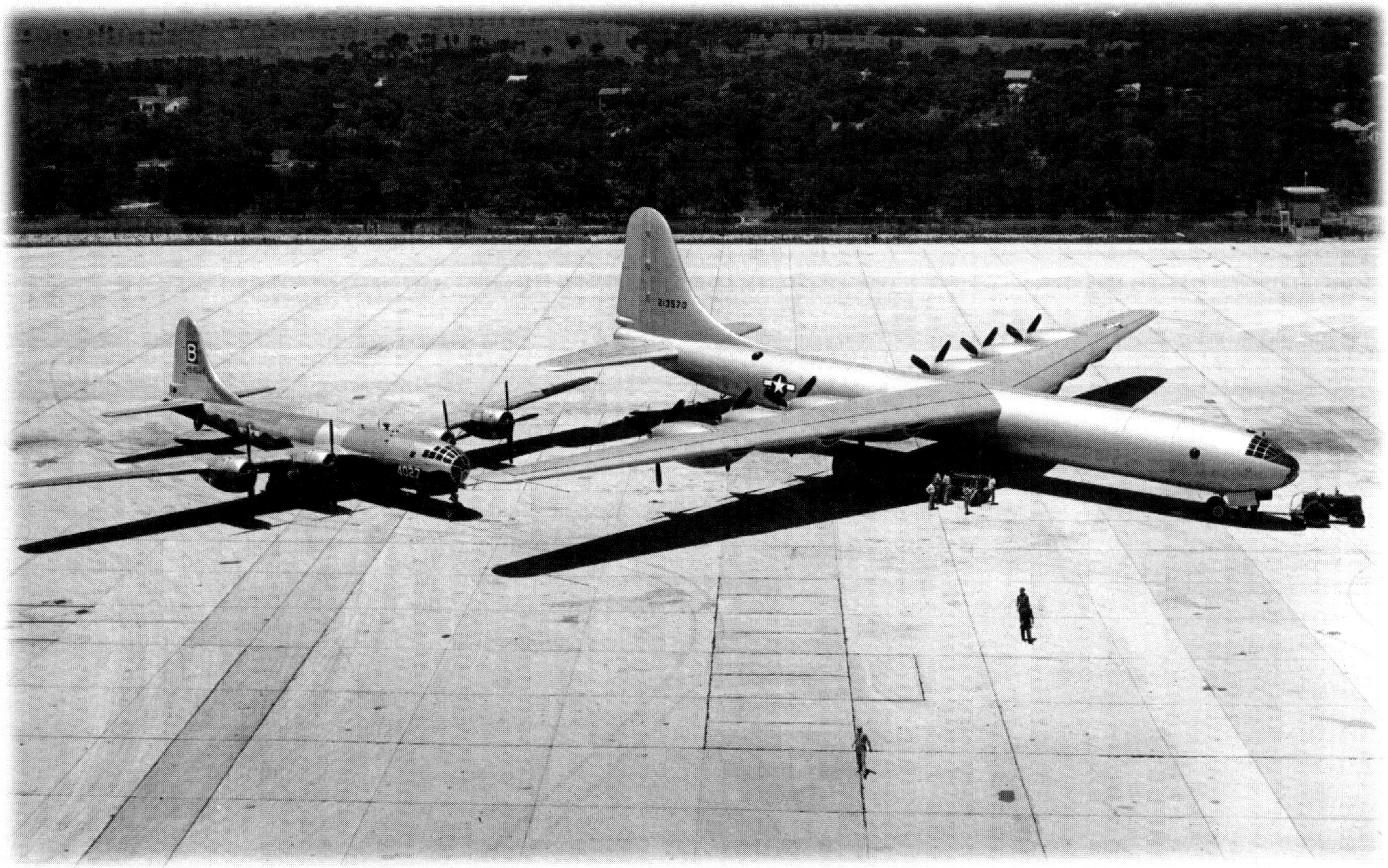

B-36 Peacemaker – As large as the B-29 was to the B-17 bomber the B-36 was even larger. The aircraft featured six Pratt and Whitney engines that generated 18,000hp. They were mounted on the trailing edge of the wing. Requiring a crew of 15, the bomber was the largest piston engine heavy bomber ever built. With a 230 foot wingspan and 163 foot length, the experimental version, XB-36, dwarfed the B-29 parked alongside it in this photograph. The B-36 was flown by the newly formed U.S. Air Force from 1949 to 1959. Known as the "Big Stick," or "Peacemaker," it was capable of delivering all of the different nuclear weapons in the U.S. arsenal from inside its two bomb bays. That ability was unique because each nuclear bomb type required different mountings. A total of 385 aircraft were built. (Consolidated-Vultee)

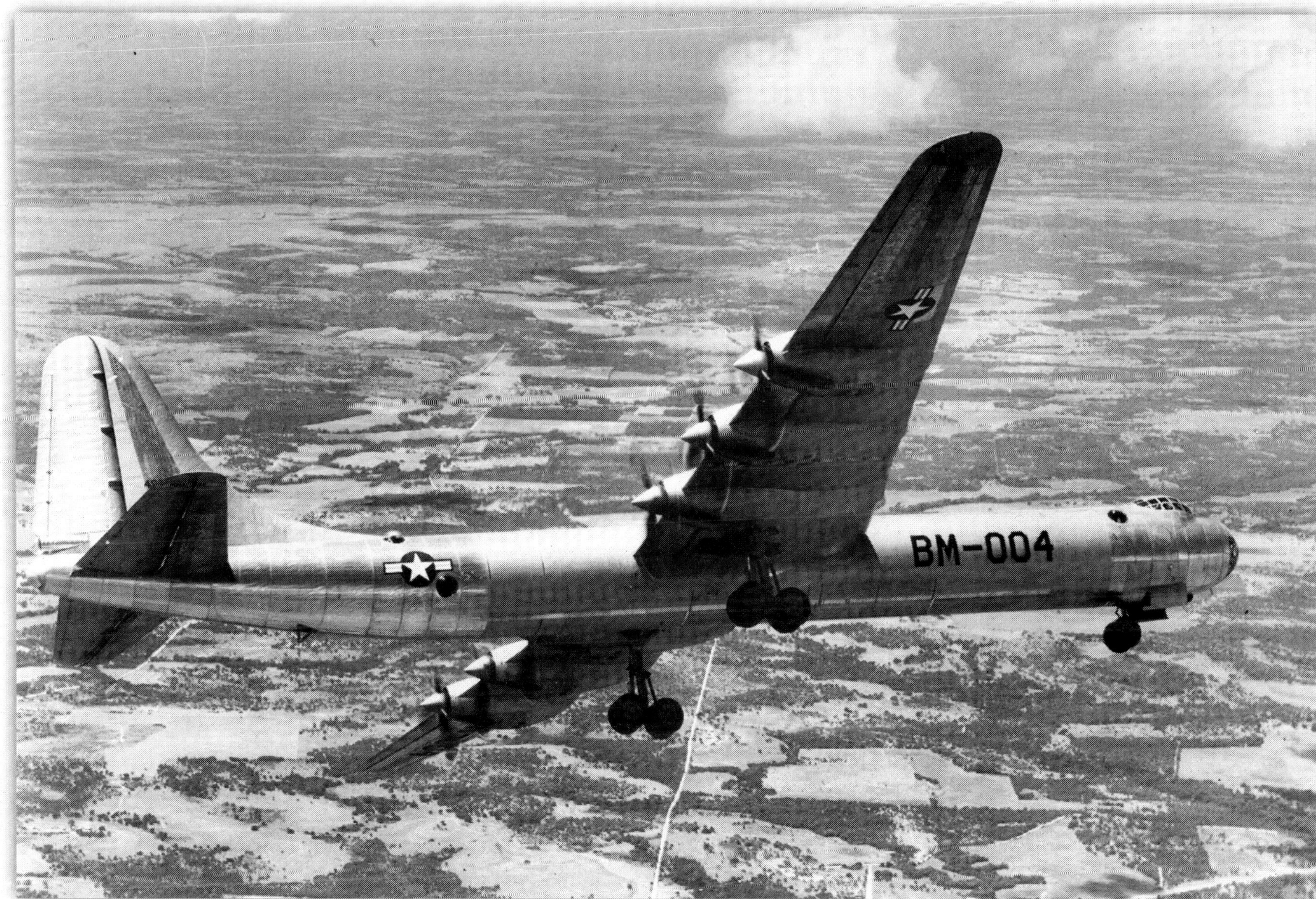

B-36 Peacemaker – The B-36 featured tricycle landing gear, like the B-24, and was the first manned bomber that could fly unrefueled with intercontinental range. Carrying one 40,000 pound A-bomb, the Peacemaker had a combat range to hit Leningrad from bases in Maine. It first flew in 1946 and became important to national defense as the Cold War escalated. This aircraft belonged to the 7th Bomb Wing, Carswell Air Force Base in Fort Worth, Texas. Although the bomber was flying with piston engines in an era of jets, no other U.S. bomber had the range and could carry a nuclear payload. (U.S. Air Force Photo, Wash., D.C.)

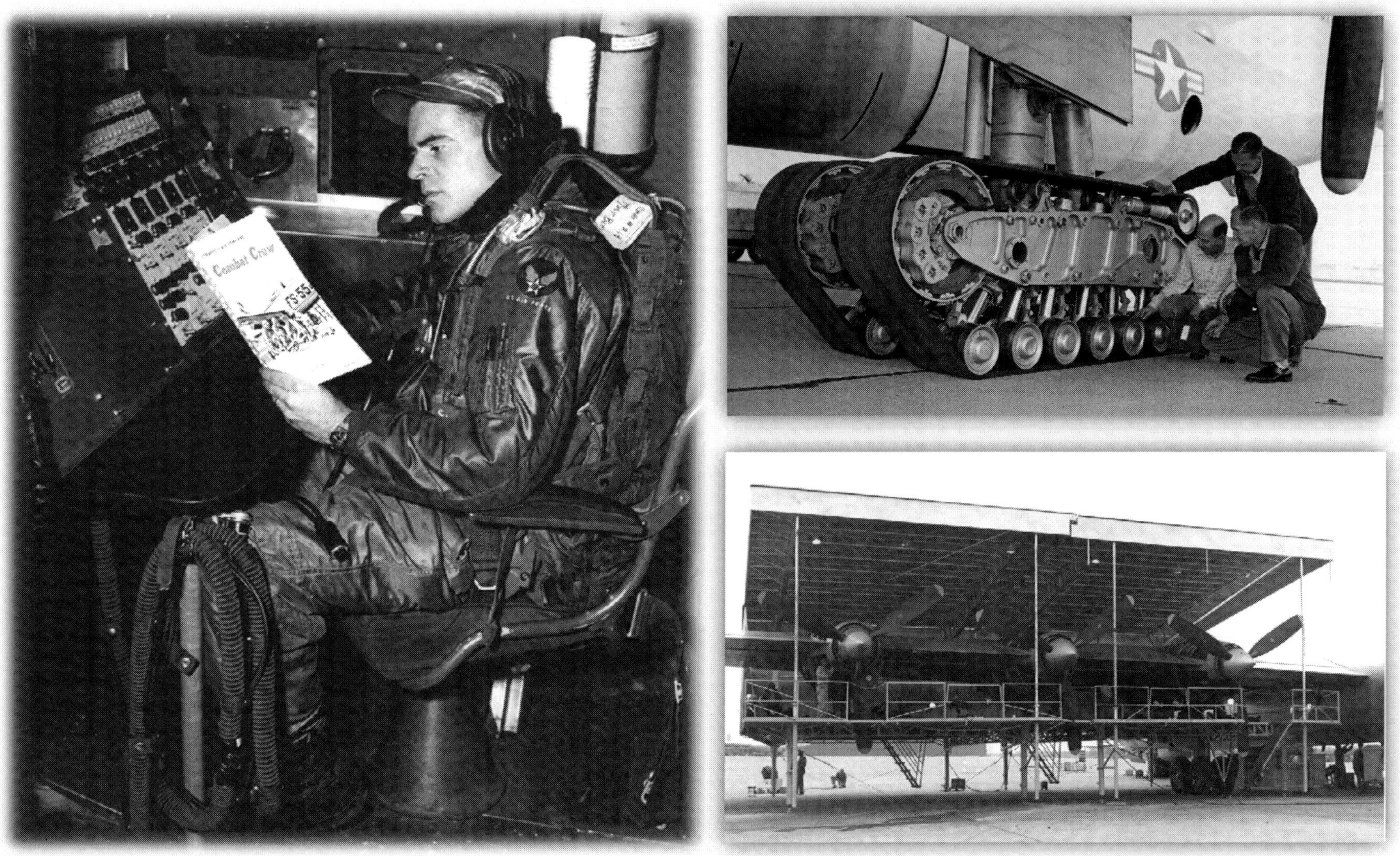

B-36 Peacemaker – Copies of the Strategic Air Command's (SAC) flying safety magazine, *Combat Crew*, were placed aboard each B-36. Technical Sergeant Carl E. Blythe was photographed reading an issue during his free moments aboard the bomber. SAC was a unique military force because it bypassed the normal chain of command. It was under the direct operational control of the President of the United States. He was the only person authorized to launch its aircraft and missiles. However, it was under the administrative control of the Air Force.[1] The March 1950 photograph (top right) shows the B-36 with experimental caterpillar-type track landing gear. It would have allowed landing on softer soils. Below right is an all-weather engine service center for the B-36 at the Fort Worth factory. (Left: U.S. Air Force and right: Consolidated-Vultee)

[1]www.strategic-air-command.com

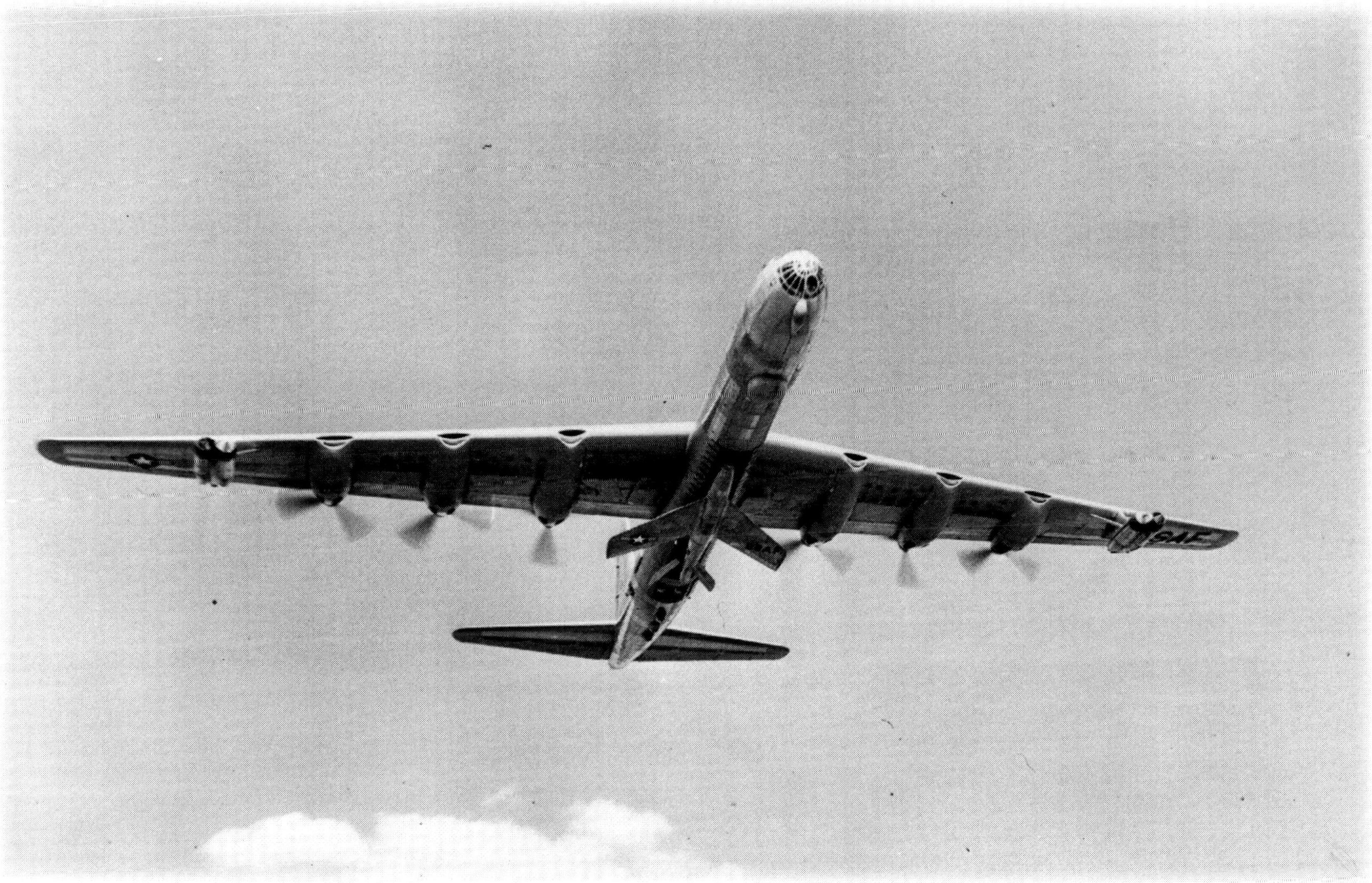

B-36 Peacemaker – In September 1950, Consolidated Vultee, the manufacturer of the B-36, announced that the aircraft was being fitted with four jet engines in addition to the standard six reciprocating engines that required 19 foot propellers. The J-47 jets seen on the end of each wing provided additional power for takeoff, improved the rate of climb, raised the service ceiling and increased the overall speed of the bomber. This photograph also shows the FICON (Fighter Conveyor) program in test. It was a program to test the feasibility of a B-36 Peacemaker bomber carrying a F-84 parasite fighter under its belly. The concept was important because no fighters had the range to protect the bomber in case of attack. (Consolidated-Vultee)

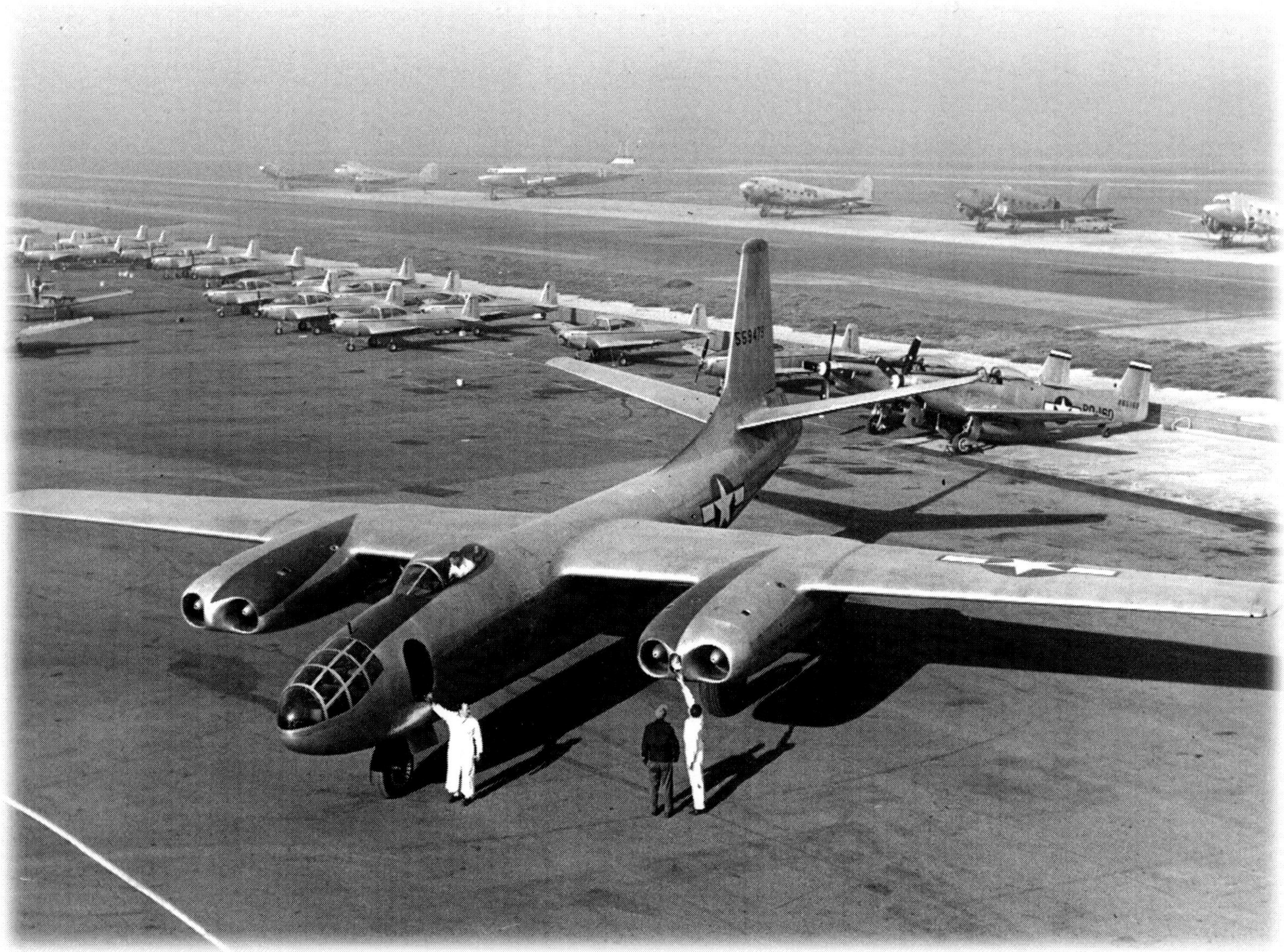

B-45 Tornado – This March 1947 photograph shows the U.S. Army Air Forces' first operational jet propulsion bomber. This XB-45 Tornado photograph provides a perspective of its size with people and nearby aircraft. Look closely and you will see a F-82 Twin Mustang fighter in the background. The B-45 was smaller than many WWII bombers with a wingspan of just 89 ½ feet and a length of 74 feet. It flew with four jet engines arranged as pairs. In production, it was known as the B-45 and used a four man crew. Post WWII bombers like the B-45 were designed to carry both conventional and atomic bombs. Carrying both types of bombs was not easily accomplished because several distinct atomic bomb types were involved. (North American Aviation)

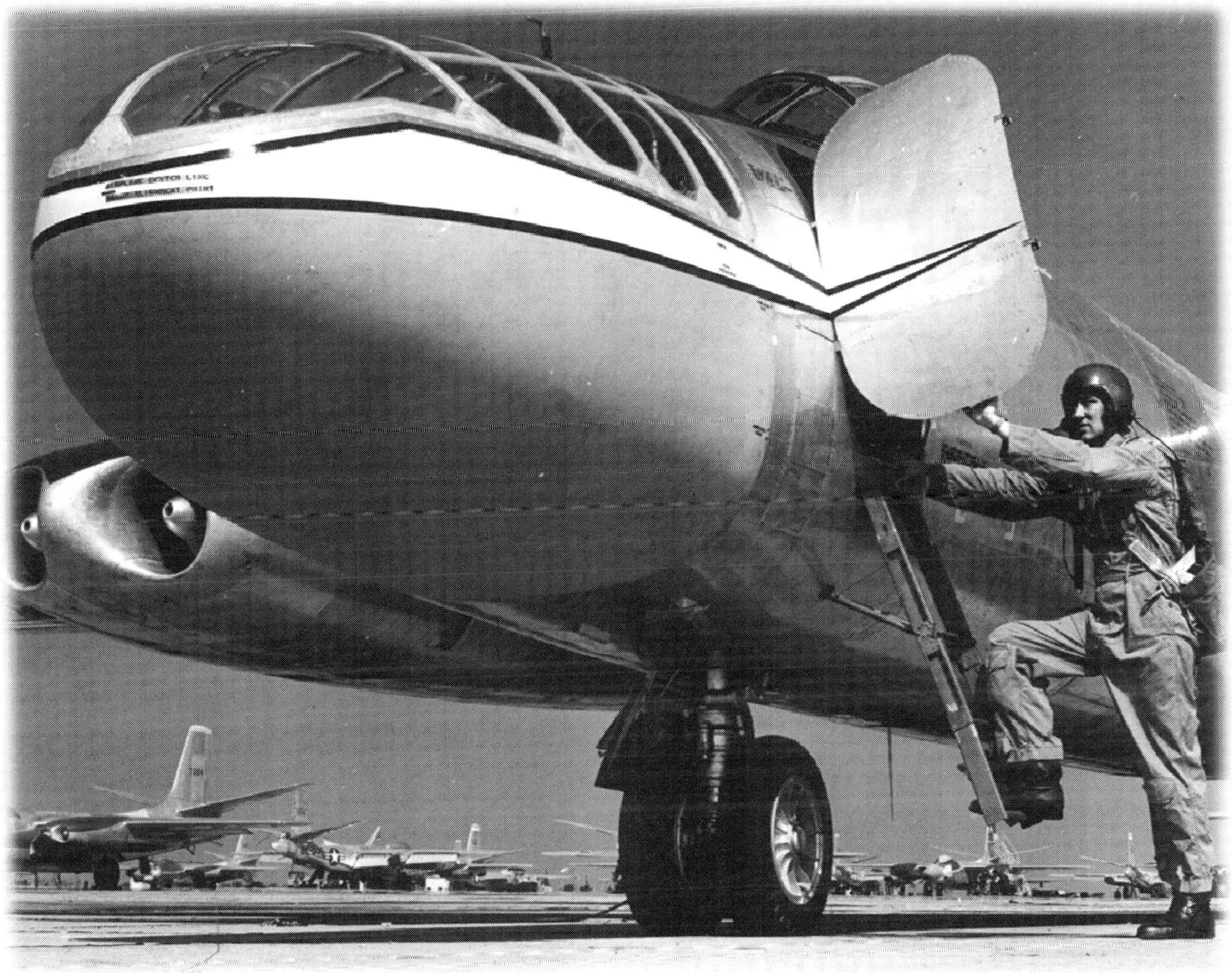

B-45 Tornado – With constant budgetary cuts slashing Air Force expenditures, the production of B-45s was reduced to a total of 142. The aircraft was important to the U.S. strategy of producing and being able to deliver large quantities of small atomic and thermonuclear weapons. A secret modification program, under the code name of Backbreaker, altered the B-45 to accommodate the strategy fully.[1] Almost on schedule, atomic capable B-45s flew to bases in the United Kingdom in May 1952. (Department of Defense) [1]www.globalsecurity.org

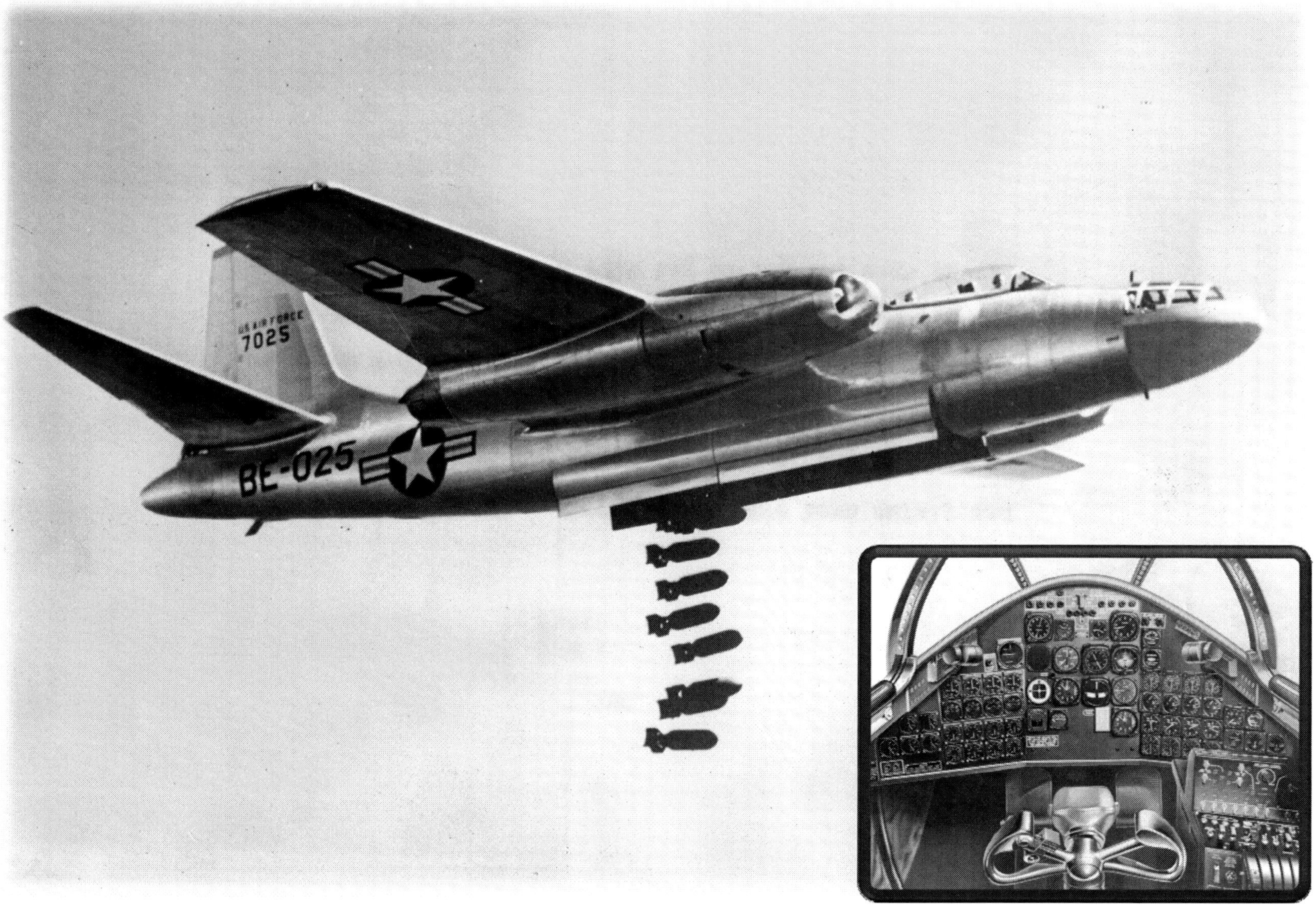

B-45 Tornado – The B-45 was in the same weight class as the WWII Boeing B-29 but flew 200 mph faster. Shown in this undated photograph, a B-45 was photographed flying at speeds over 500 mph while it dropped 500 pound bombs safely and accurately for the first time. The B-45 also flew in a reconnaissance configuration known as the RB-45C. The Strategic Air Command directed the use of reconnaissance Tornados as they performed classified, deep penetration photographic intelligence missions over many cold war communist countries including Korea. The reconnaissance version of the B-45 became the forerunner of the U-2 and SR-71 surveillance aircraft.[1] (Both photographs by North American Aviation) [1]www.globalsecurity.org

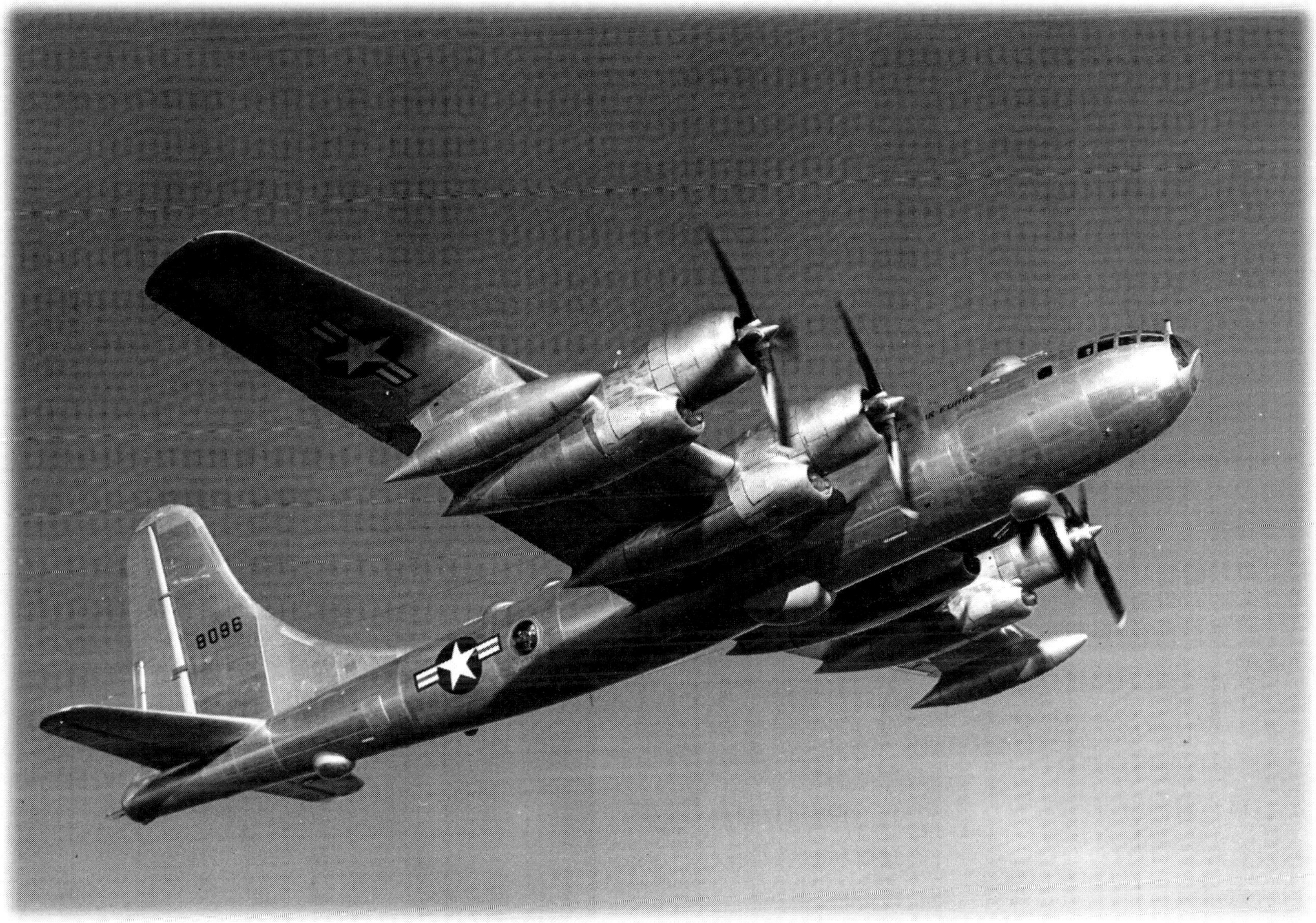

B-50 Superfortress – The Cold War B-50 could almost pass as a modified B-29 WWII bomber on first glance. Although it maintained a strong similarity in physical appearance to the B-29, its components were 75% reengineered. Flying with a crew of ten and the capability of carrying 20,000 pounds of ordnance, the B-50 provided the Strategic Air Command with a powerful nuclear deterrent in the face of the expanding Cold War. The aircraft entered service in June 1948 and was phased out in October 1955. (Boeing Airplane Company)

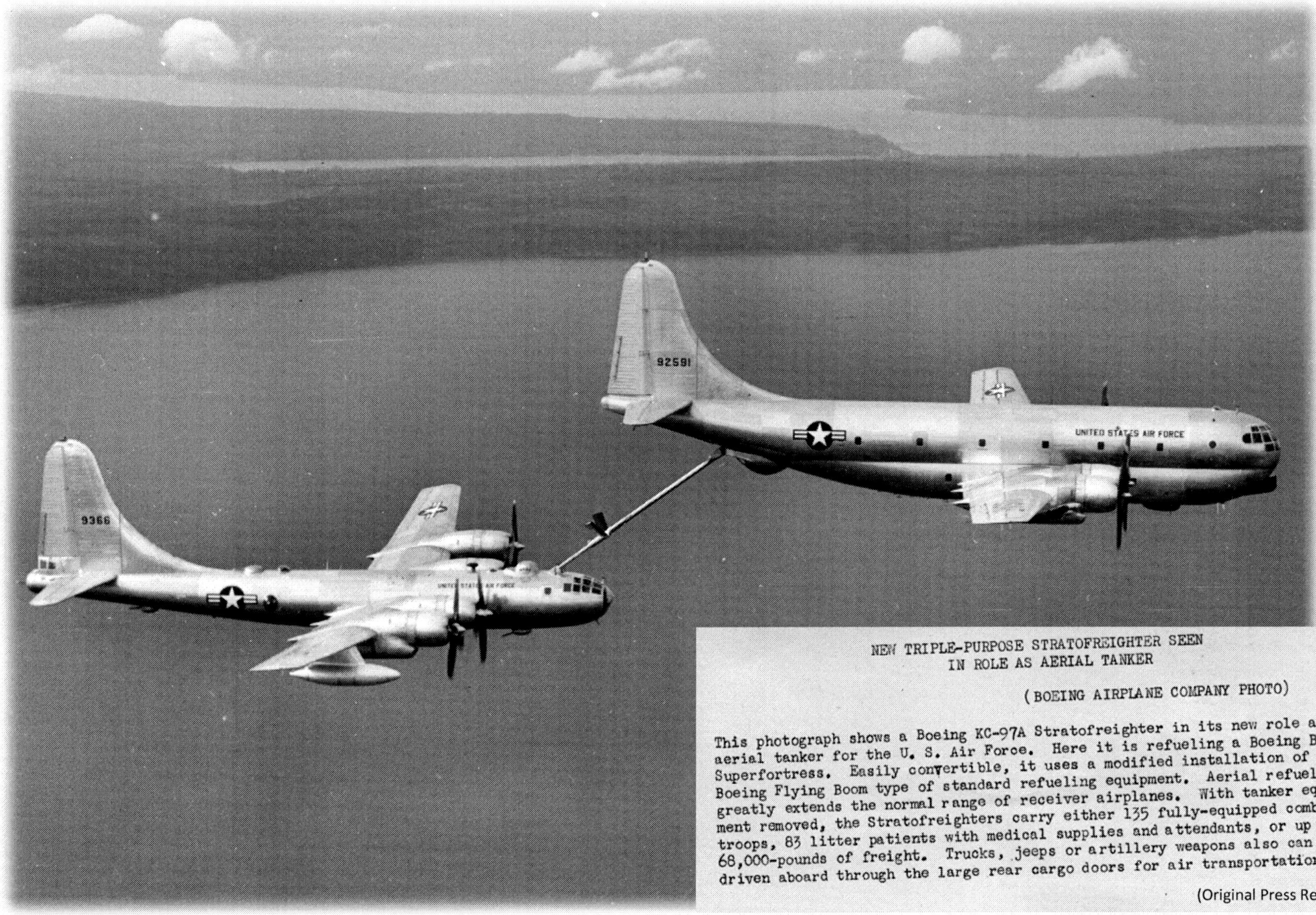

NEW TRIPLE-PURPOSE STRATOFREIGHTER SEEN
IN ROLE AS AERIAL TANKER

(BOEING AIRPLANE COMPANY PHOTO)

This photograph shows a Boeing KC-97A Stratofreighter in its new role as an aerial tanker for the U. S. Air Force. Here it is refueling a Boeing B-50D Superfortress. Easily convertible, it uses a modified installation of the Boeing Flying Boom type of standard refueling equipment. Aerial refueling greatly extends the normal range of receiver airplanes. With tanker equipment removed, the Stratofreighters carry either 135 fully-equipped combat troops, 83 litter patients with medical supplies and attendants, or up to 68,000-pounds of freight. Trucks, jeeps or artillery weapons also can be driven aboard through the large rear cargo doors for air transportation.

(Original Press Release)

B-50 Superfortress – The early Cold War did not feature intercontinental ballistic missiles which kept bomber range and speed a crucial deterring factor to war. Seen above with the B-50, left, is a KC-97A Stratofreighter aerial tanker. The B-50 had an operational range of 2,394 miles and could be refueled in the air. The advent of rudimentary aerial refueling began in the 1920s[1] but by the fall of 1949, B-50s were refueled much like today. With the use of KB-29M tankers, B-50 crews were flying training missions with distances over 5,000 nautical miles. (Boeing Airplane Company) [1]See Memo Report, Subject: Refueling in Flight, Capt D. L. Yeager, Army Air Forces Engineering Division, 25 June 1943, *Case History Supporting Documents.*

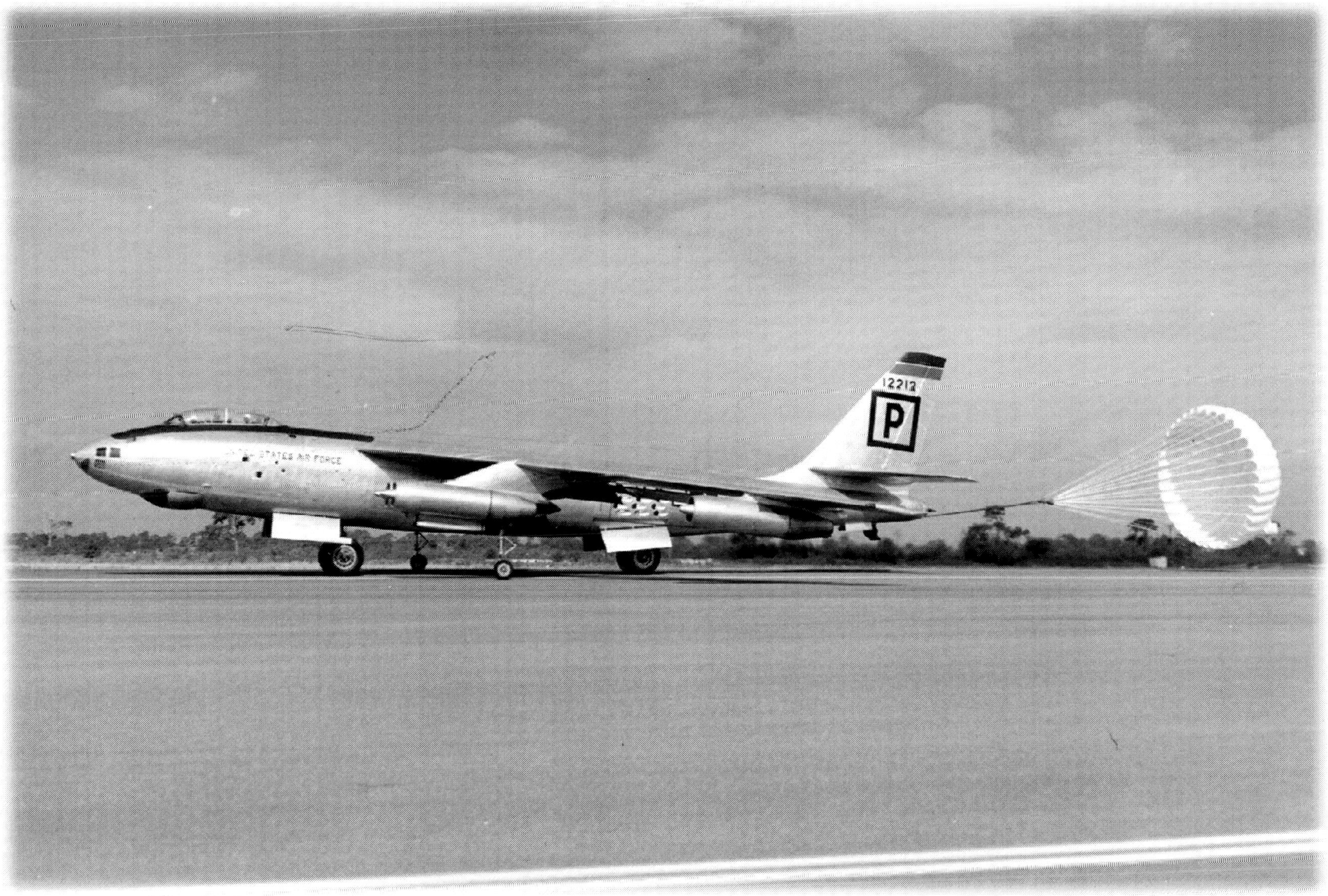

B-47 Stratojet – The B-47 Stratojet flew as the U.S. Air Force front line nuclear bomber until the B-52 was available in the late 1950s. The first prototype of the B-47 bomber flew in December 1947 with six Allison turbojet engines hung beneath the swept-back wings. The engine power and design gave the bomber the speed and maneuverability equal to its contemporary fighters but the aircraft required a high landing speed and an excessively long distance to stop as a result of its design. A three man crew controlled the aircraft that had a wing span of 116 feet and was 107 feet long. (U.S. Air Force)

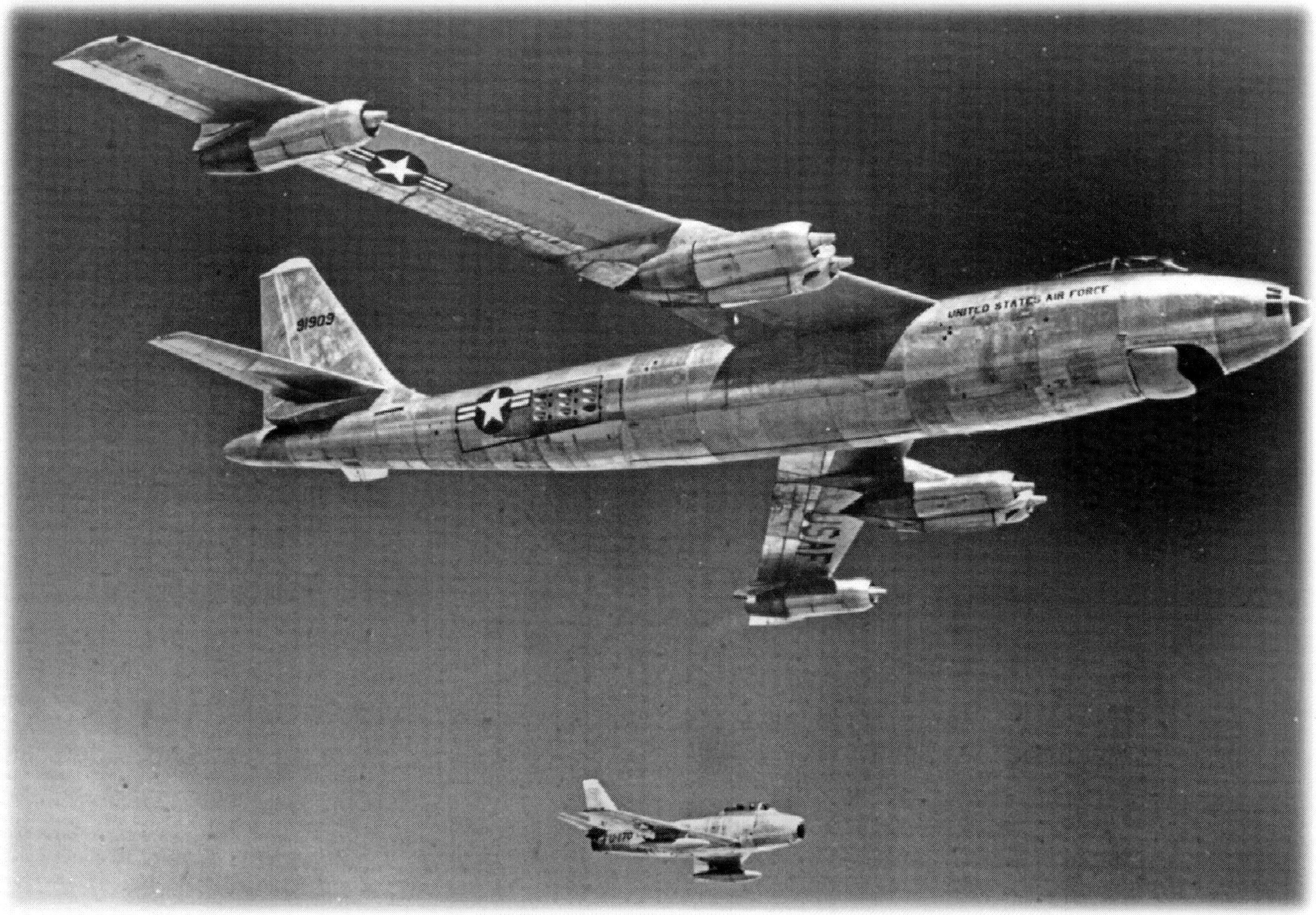

B-47 Stratojet – This B-47 bomber was photographed along with a F-86 fighter escort. Designed to replace the B-50, the B-47 entered production in 1952 and was produced until 1956. A total of 2,039 were built, more than any other U.S. bomber manufactured under peacetime conditions.[1] It was capable of a 3,500 mile range. The bomber was close to having been canceled because its capabilities exceeded those specified for a medium bomber and fell short of those of a heavy bomber.[2] Intercontinental missiles were not yet operational and a fast long range aircraft was the only nuclear strike vehicle available. (General Electric)

[1&2]http://reference.findtarget.com/search/Guy%20M.%20Townsend/

B-52

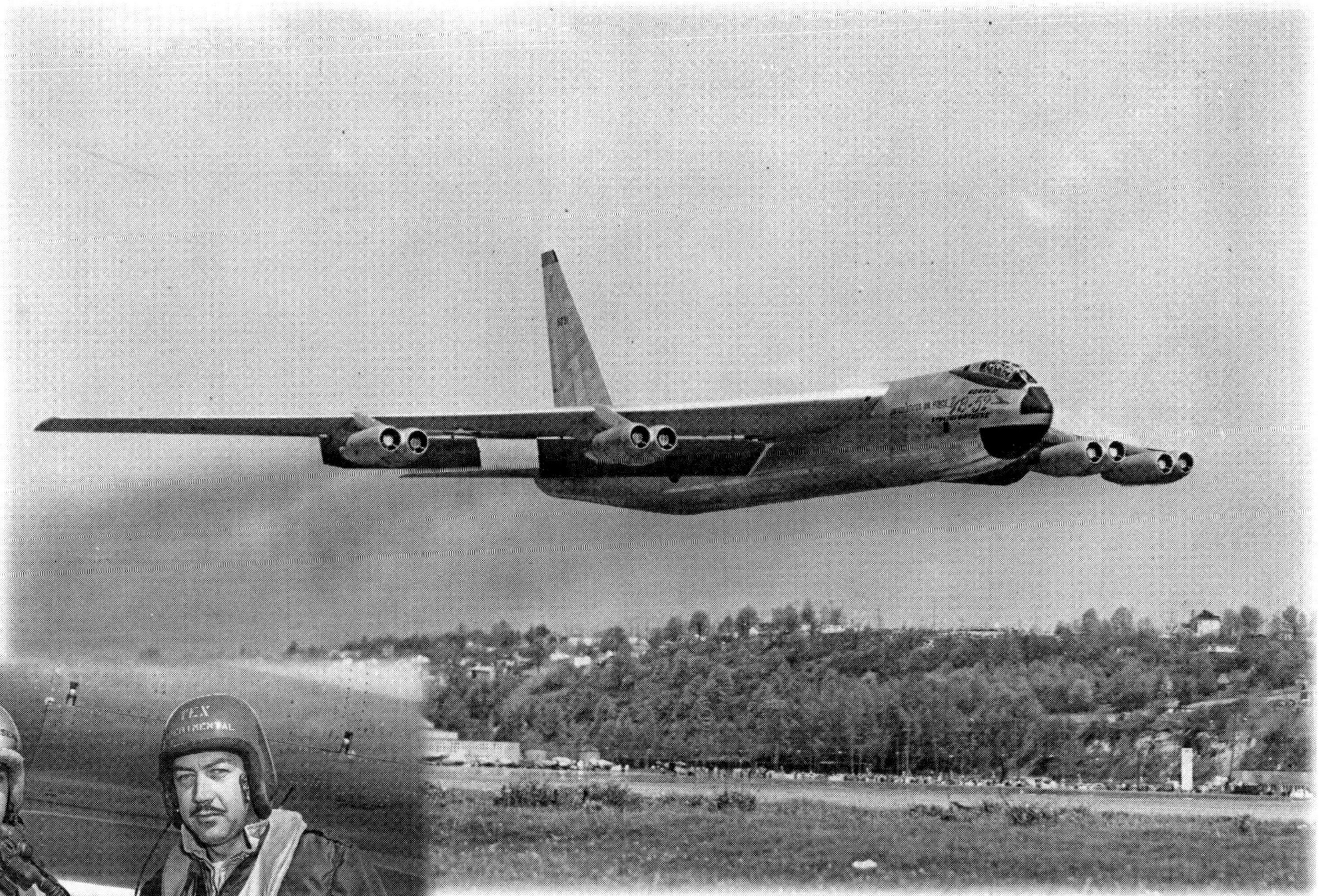

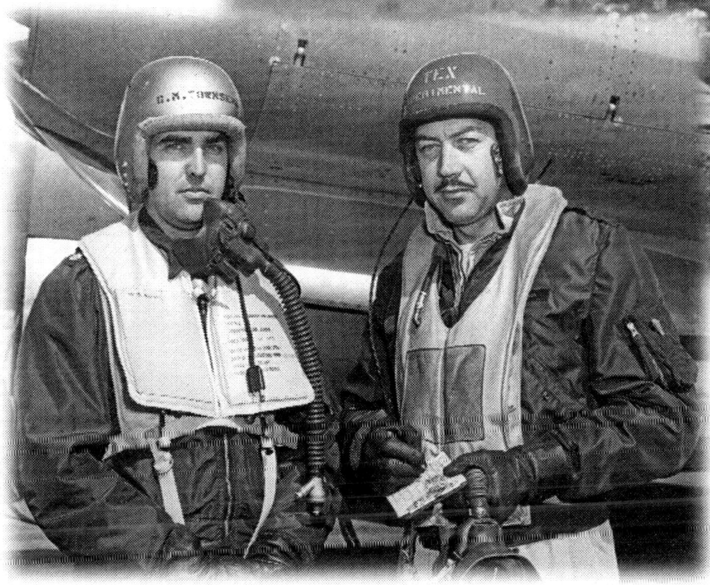

B-52 Stratofortress – Seen here, the YB-52, the test version of the B-52, as she had just begun her maiden flight. Its first flight was two hours and 51 minutes long. Pictured to the left are the pilot and copilot who were at the controls of the aircraft on this flight. Left to right are Lieutenant Colonel Guy M. Townsend, copilot, and A.M. Tex Johnston, pilot.

The B-52 first flew on April 15, 1952 and entered active service in 1955. It has been the preeminent American heavy bomber ever since. The aircraft was designed to replace the B-36 heavy bomber. (Both photographs by Boeing Airplane Company)

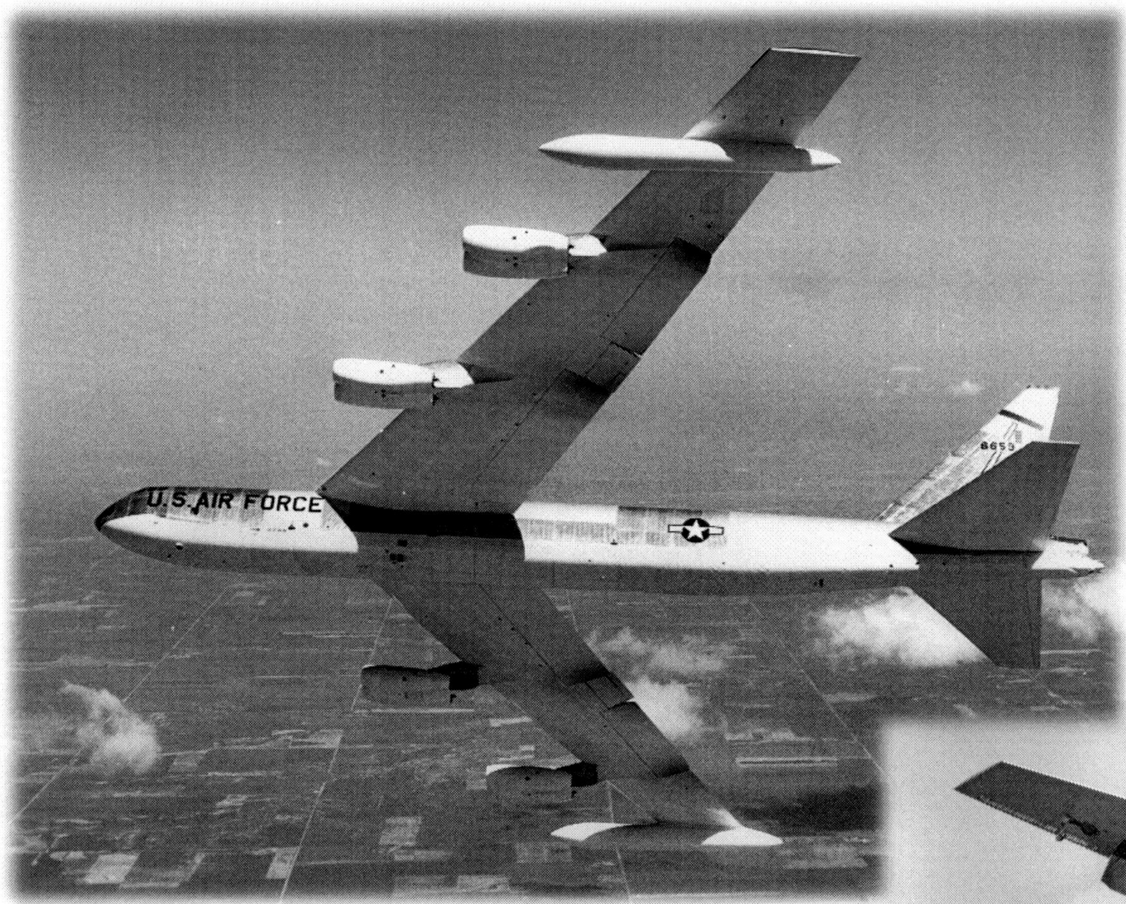

B-52 Stratofortress

A production B-52 was photographed in May 1957 during a routine flight over the U.S. Midwest. The 400,000 pound bomber carried a crew of six and had a range of over 6,000 miles. Flying above 50,000 feet at more than 650 mph, the aircraft presented a formidable image of deterrence during the Cold War. With a wingspan of just 185 feet, a length of 153 feet and over 40 feet tall, it was smaller than the B-36 but taller.

The two-story cockpit presented a unique climate control problem while parked on a tarmac – the pilot's cockpit was heated by sunlight while the observer and the navigator on the bottom deck sat on the ice-cold floor.[1] (Boeing Airplane Company)

[1]Knaack, Marcelle Size. *Post-World War II Bombers, 1945–1973.* Washington, DC: Office of Air Force History, 1988. ISBN 0-16-002260-6.

B-52 Stratofortress

The B-52 photograph to the right shows the unusual eight wheel main landing gear of the YB-52. The aircraft was so heavy many runways crumbled under its mass. The picture also shows the outrigger wheels stationed near each wing tip. The early B-52 jet bombers used four double Pratt & Whitney J-57 jet pods. This photograph was released on May 13, 1952. (Boeing Airplane Company)

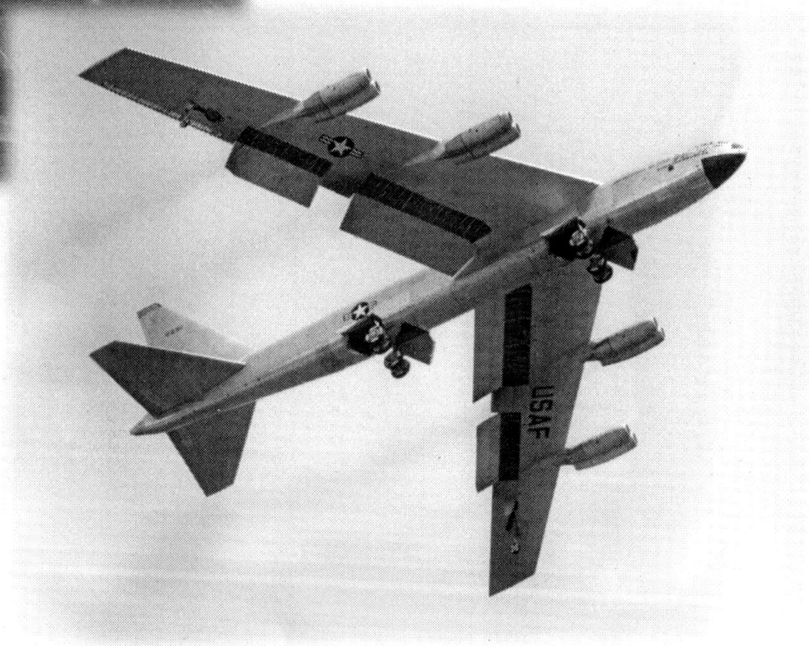

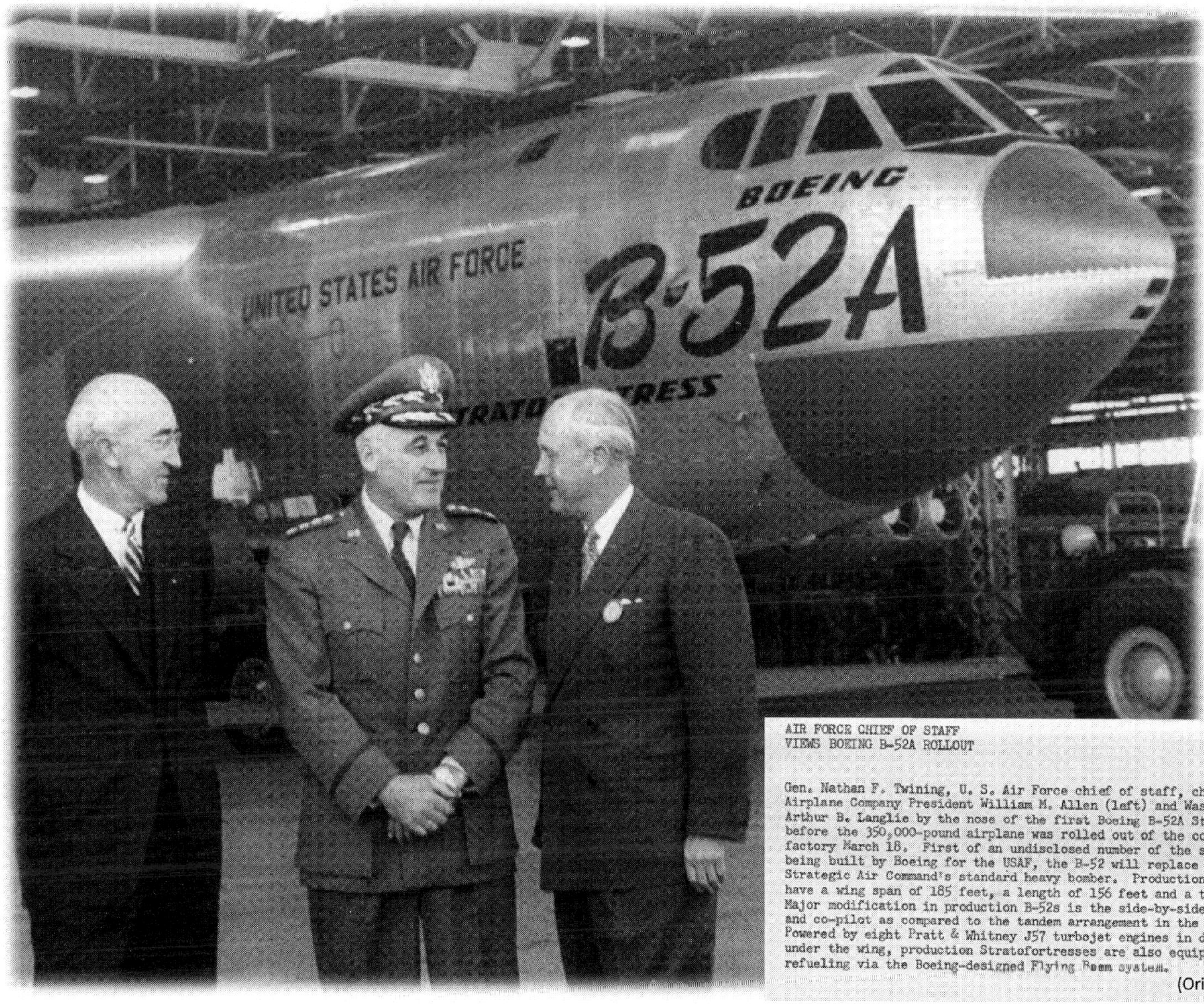

AIR FORCE CHIEF OF STAFF
VIEWS BOEING B-52A ROLLOUT

Gen. Nathan F. Twining, U. S. Air Force chief of staff, chats with Boeing Airplane Company President William M. Allen (left) and Washington's Governor Arthur B. Langlie by the nose of the first Boeing B-52A Stratofortress just before the 350,000-pound airplane was rolled out of the company's Seattle factory March 18. First of an undisclosed number of the swept-wing bombers being built by Boeing for the USAF, the B-52 will replace the B-36 as the Strategic Air Command's standard heavy bomber. Production Stratofortresses have a wing span of 185 feet, a length of 156 feet and a tail height of 48 feet. Major modification in production B-52s is the side-by-side seating for pilot and co-pilot as compared to the tandem arrangement in the experimental models. Powered by eight Pratt & Whitney J57 turbojet engines in double pods slung under the wing, production Stratofortresses are also equipped for aerial refueling via the Boeing-designed Flying Boom system.

(Original Press Release)

B-52 Stratofortress – Only three B-52A models were built and they never entered service with the U.S. Air Force. Two were used on additional test beds for improvements and other needs. The third B-52A was modified and re-designated NB-52A in 1959 to carry the X-15 rocket-powered aircraft/spaceplane. The B model version entered service on June 29, 1955. Throughout the Cold War, B 52s performed airborne patrols loitering at high altitude outside the Soviet Union to provide rapid first strike or retaliation capability in case of nuclear war. (Boeing Airplane Company)

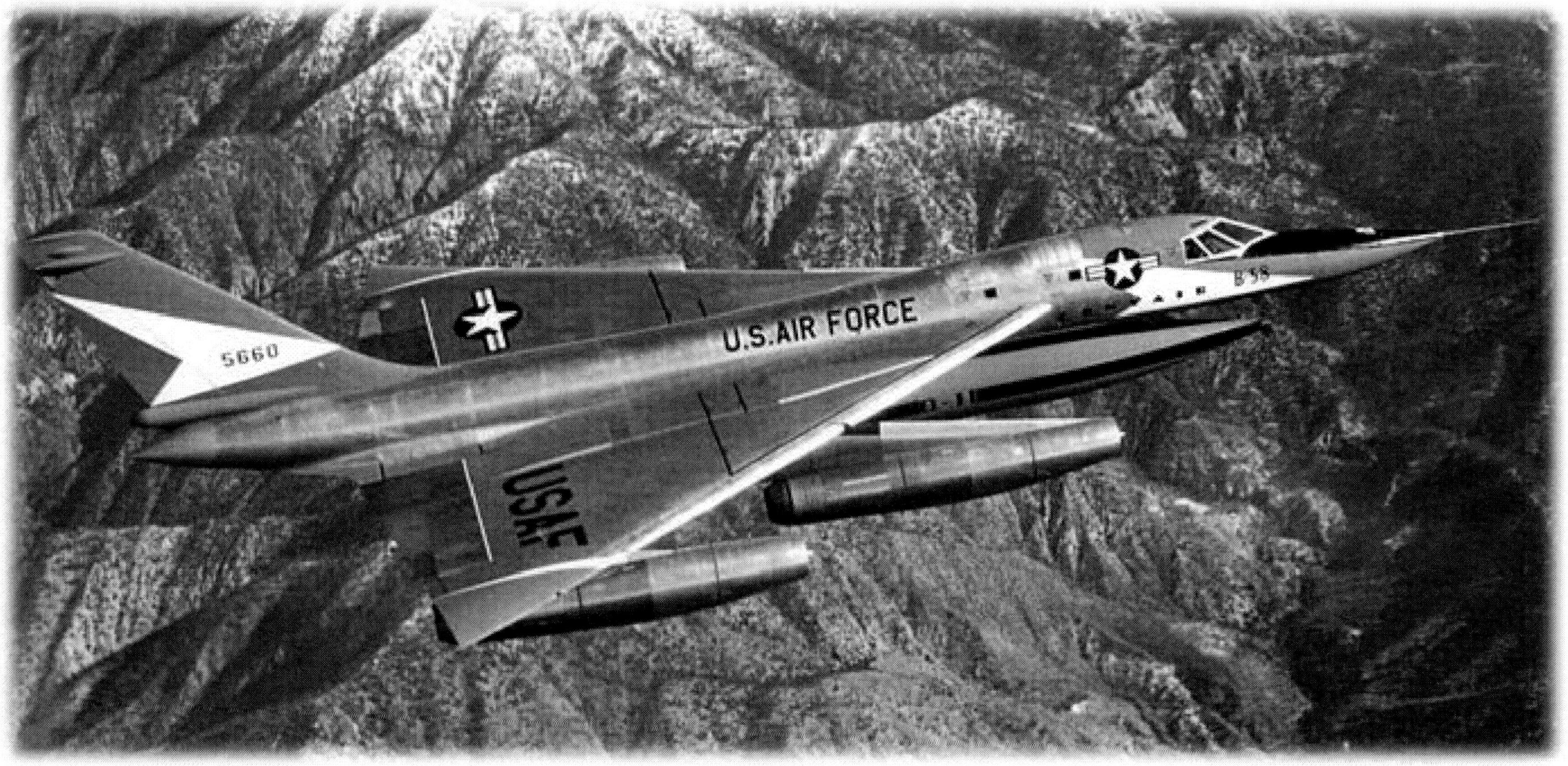

B-58 Hustler – The B-58 was the first U.S. Air Force supersonic bomber, making its initial flight on November 11, 1956 and flying supersonically on December 30, 1956. Much smaller than contemporary bombers, it was almost 97 feet long with a narrow wingspan of just 56 feet 9 inches. It could carry 19,450 pounds of ordnance, more than twice the capacity of the WWII B-17 bombers. The three crew members sat in tandem escape capsules in case they needed to eject. This was important because the aircraft was capable of Mac 2 (1,319 mph) and cruised at 650 mph.

Surface to air missile defense shortened the useful life of this bomber, forcing it to fly low level penetration routes with a limited range of just over 1,700 miles. There were 116 B-58s built of which 86 were deployed as official operating inventory. They were operational in the Strategic Air Command (SAC) Cold War maneuvers between 1960 and 1970. (U.S. Air Force)

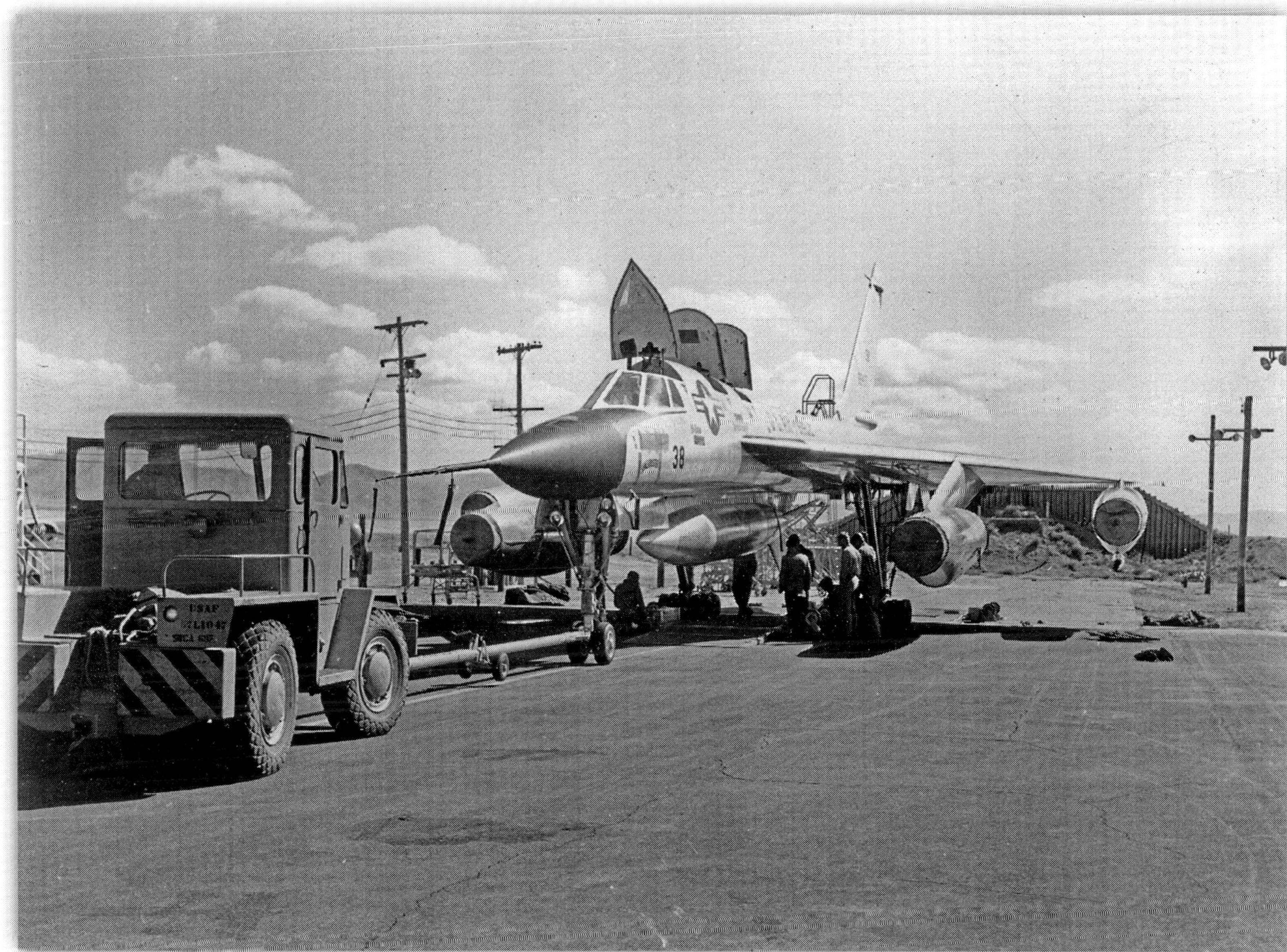

B-58 Hustler – This B-58 was being readied at the Kirtland Air Force Base for determination of gross weight and center-of-gravity positioning. The vehicle at the left moved the aircraft into position using a tow bar attached to the B-58 nose gear. Empty, the aircraft weighed 55,560 pounds and its maximum takeoff weight was 176,890 pounds. (Baldwin-Lima Hamilton Corporation)

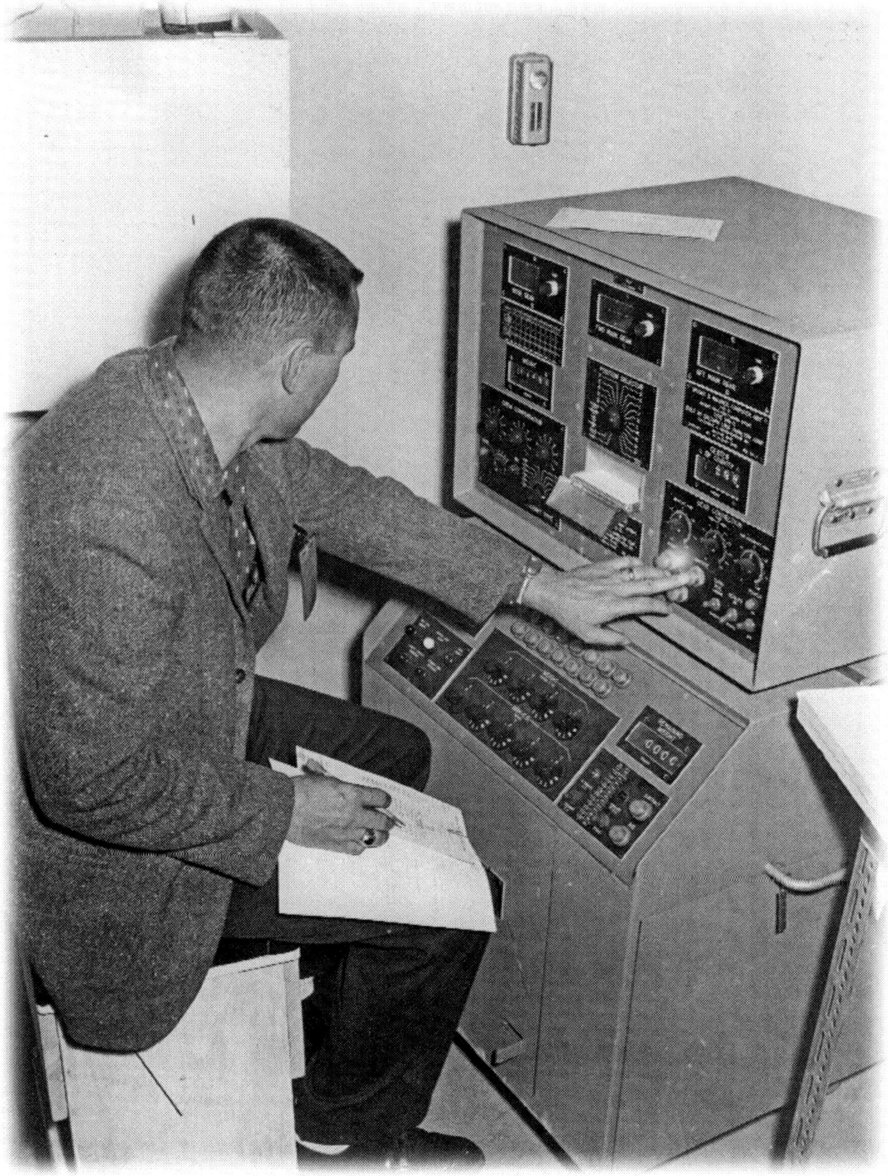

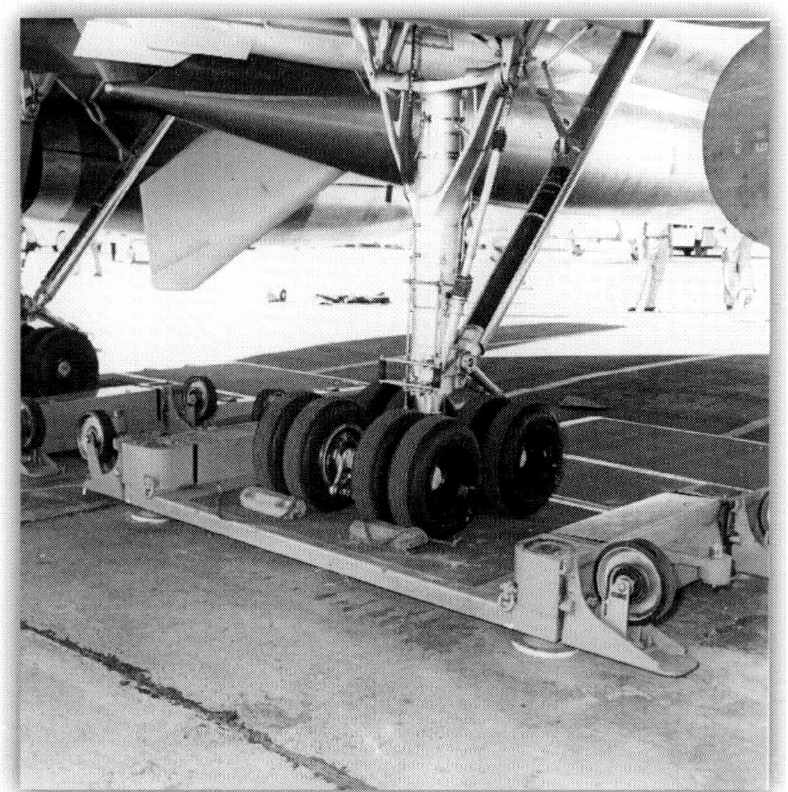

B-58 Hustler – Senior Flight Test Engineer O.D. Lively, from General Dynamics Corporation, Fort Worth, TX, was photographed while operating the B-58 Computer-Planner Console during a B-58 weigh in. The B-58 flew in the nation's aerial arsenal for only ten years yet helped set the direction for future bomber design and broke many barriers to supersonic flight. (Both photographs by Baldwin-Lima Hamilton Corporation)

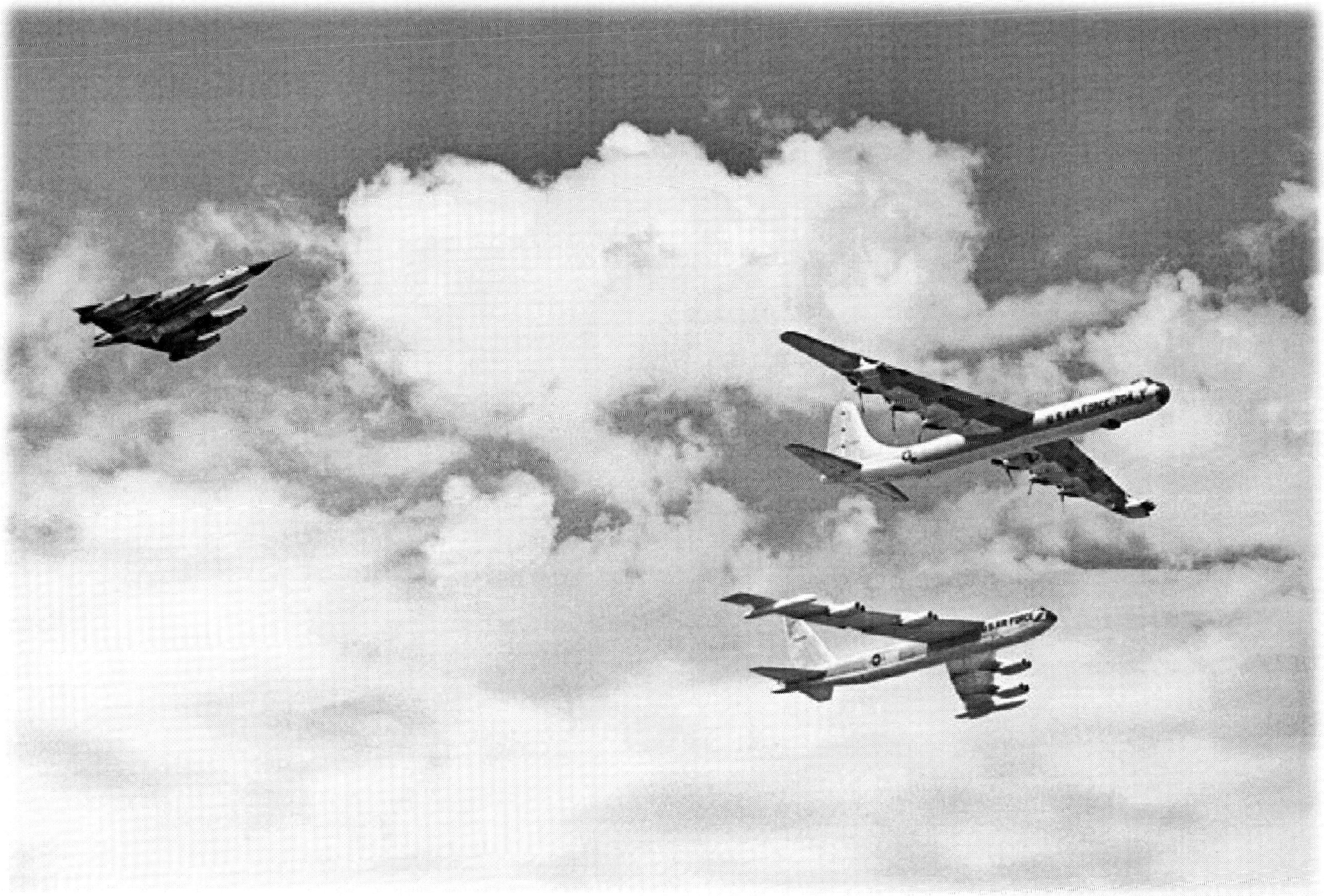

B-58 Hustler – The B-58, left, flew alongside its two predecessors the B-52 bomber, lower right, and the B-36, upper right. The B-36 was being phased out of service about the same time that the B-58 was entering service. The B-58 was retired in 1970 while the B-52 remains in the U.S. air arsenal. (U.S. Air Force)

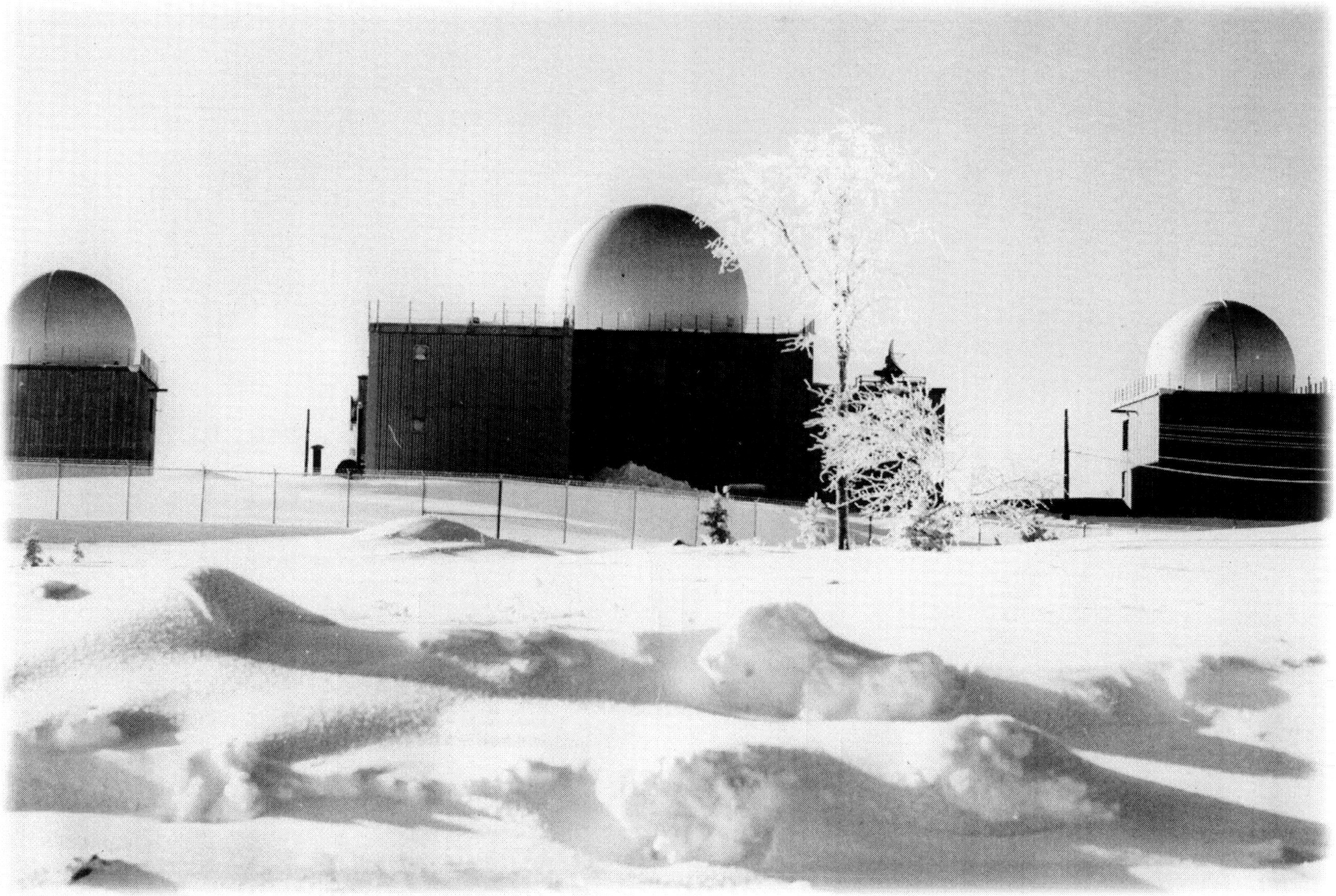

The Cold War – As bomber technology increased during the Cold War, NATO and Russia tested their opponent's early warning systems. Bombers who normally flew just outside of a country's airspace would turn toward the enemy border to test how quickly they would obtain a response. Slowly, bombers partnered with intercontinental missiles as a first strike force increasing the importance of an early warning system. These domes housed the Pine Tree Line radar outpost near Quebec, Canada, which was the most southerly of NATO's radar defense fence. (HQ Air Defense Command)

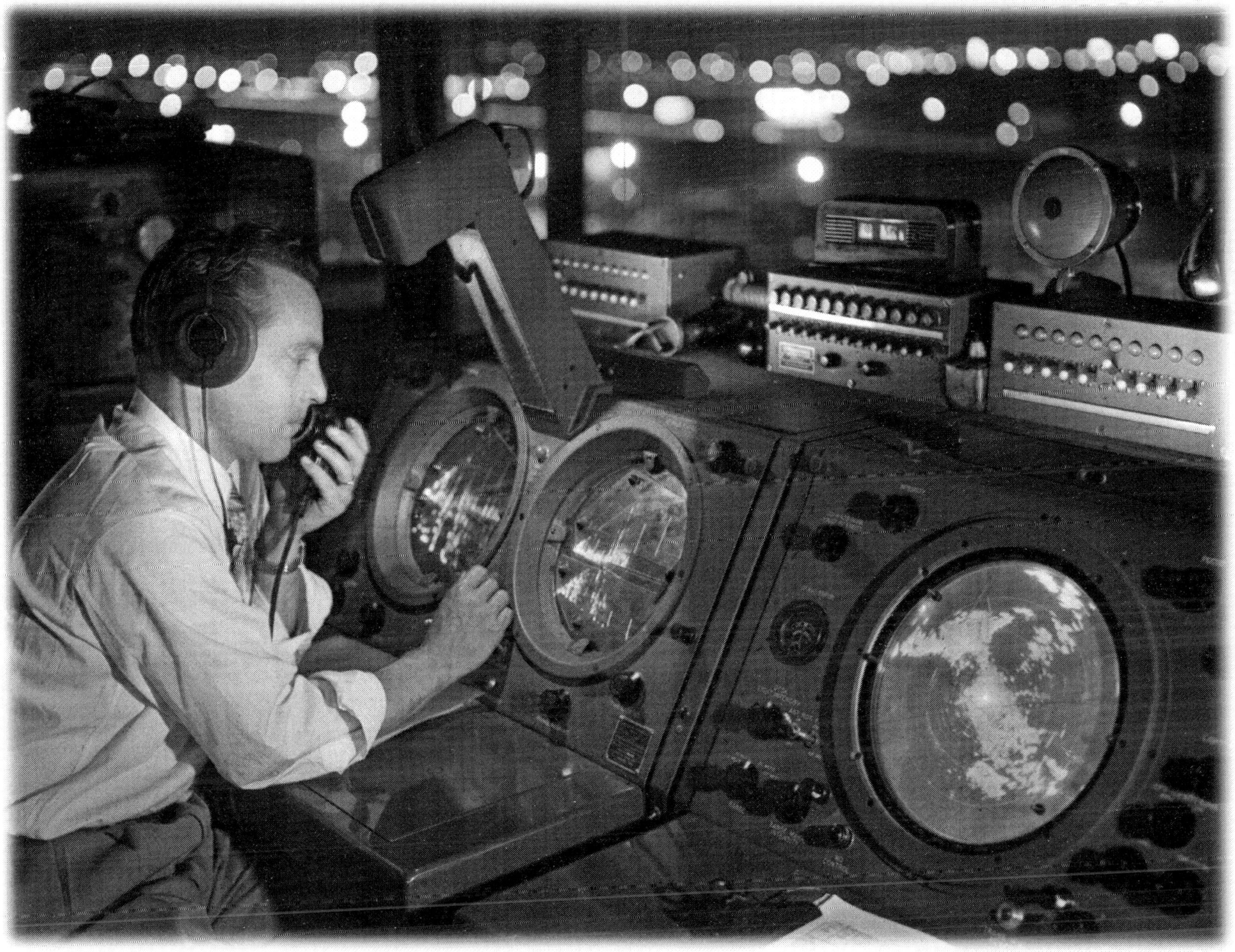

The Cold War – These radar units were on a peaceful mission tracking airline traffic at the Los Angeles Airport but they are not unlike the screens that dedicated NATO servicemen watched for hours on end, waiting for the enemy to breach NATO airspace. (William Eccles)

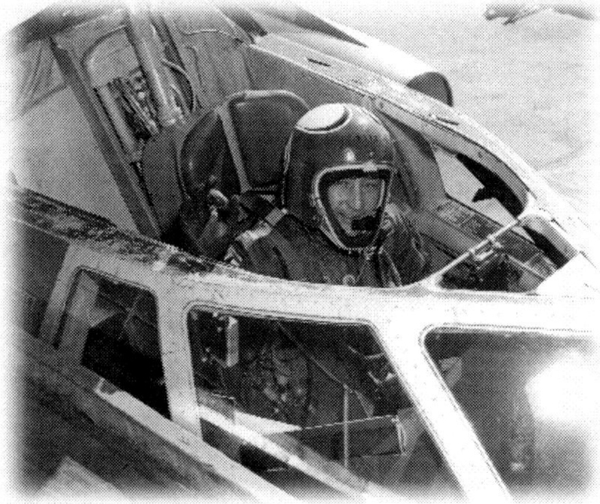

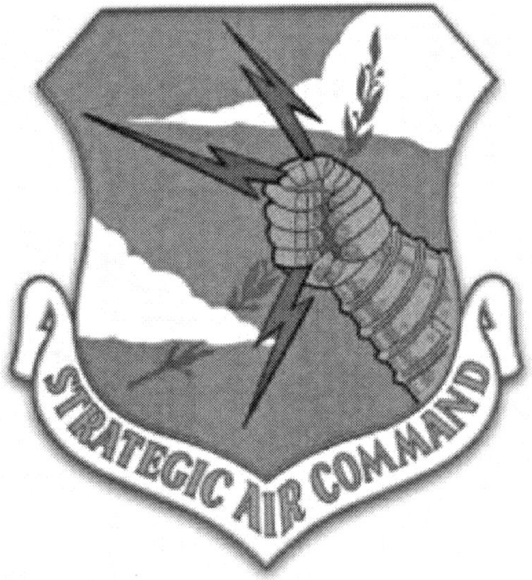

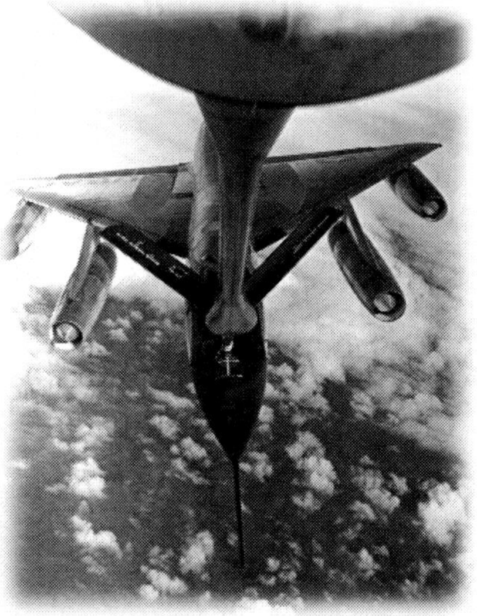

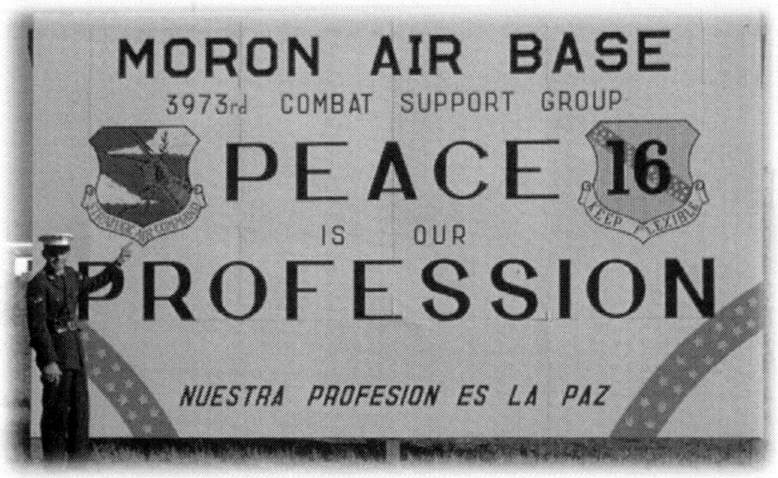

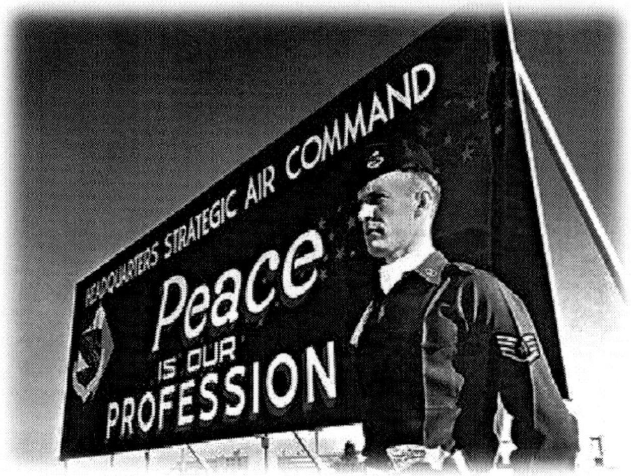

A Tribute to the Strategic Air Command – During the Cold War, from 1946 until 1992, SAC managed America's proverbial "Big Stick" using the combined airpower of bombers and Intercontinental Ballistic Missiles (ICBMs) as a deterrent to a third world war. The rules of engagement had changed; the world could not afford a win/lose strategy any longer since weapons had become capable of mass destruction. The risks that SAC managed are part of history now. The stakes they managed were immense. There were no rulebooks.

The Cold War has subsided and détente versus annihilation has become more appealing to world leaders. In 1992, SAC merged with other U.S. Air Force organizations and established the Air Combat Command (ACC) which now manages America's airpower.[1] We thank the thousands of people that worked under incalculable stresses to deliver on SAC's promise to America – Peace. (Patch: Strategic Air Command, Photographs top right and left, U.S. Air Force, Photograph left, Heinzel Barry & www.3973cds.com, Photograph right: Strategic Air Command) [1]U.S. Air Force

American Airpower: Bombers 1935-1960 a conclusion –

During the brief twenty-five year history covered in this photographic record, bomber technology changed dramatically. The twenty-five year evolution developed aircraft that flew faster, farther and carried more destructive bomb loads. In WWII, formations of bombers were required to destroy a target. The aircraft and crews flew dangerous missions that often suffered high casualty rates. As the war progressed, bomber and supporting fighter tactics improved which also improved survival rates.

The Korean War was transitional in regard to bombers. Many WWII bombers were redeployed and their crews fought similar battles. Simultaneously, aircraft with jet engines were being deployed during the war. Jet bombers, like the B-45, began to be used in surveillance of the enemy.

As the Cold War escalated, both before and after Korea, new bombers that been in development in the late 1940s were being deployed. At first it was the Peacemaker, a propeller driven aircraft in the age of jets that had enough range to reach a potential enemy. Soon long range jet bombers like the B-52 entered the arsenal. They were supported by the shorter range, but faster, B-58.

As the 1960s progressed, bomber technology and missile technologies began to intertwine. Aircraft began to carry missiles that could be launched from the air and deliver explosives hundreds of miles from the launch site. Intercontinental missiles and their silos were deployed around the world integrating the defensive and offensive military first strike roles across the technologies.

Bombers continue their evolution and today their job remains the same as the B-10 bomber of 1935, to drop explosives from the air.

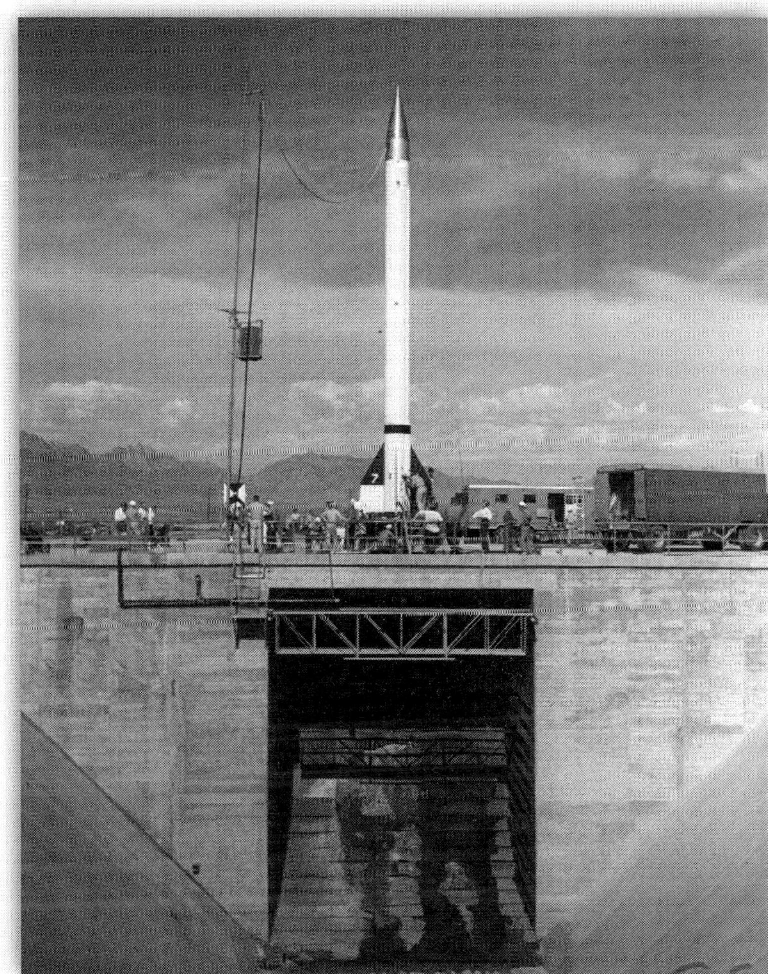

Martin Viking rocket – This photograph was taken while the Glenn L. Martin Company was providing a final rocket check prior to Viking 7's record breaking flight of 136 miles. The rocket was launched as part of a U.S. Navy Research Laboratory experiment on August 7, 1951. The concrete pit below the rocket was lined with steel plates to absorb the intense heat of the rocket engines and was used for static firings of the rocket. (Dick Darrow & Glenn L. Martin Company)

Index

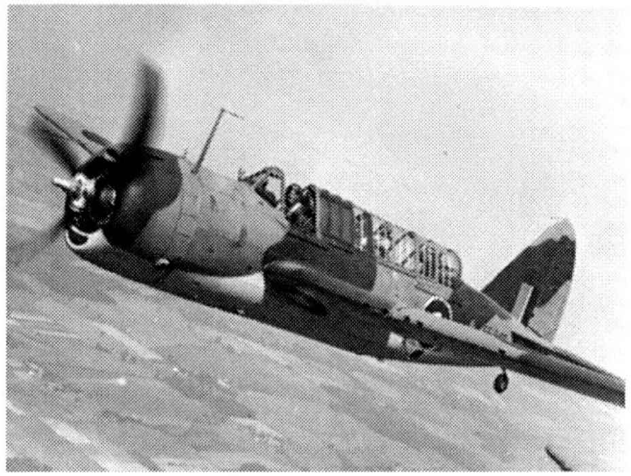
Buccaneer (Brewster)

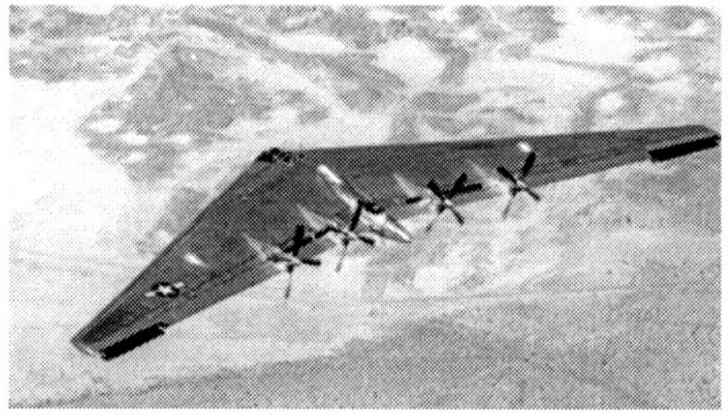
YB-35 (Northrop)

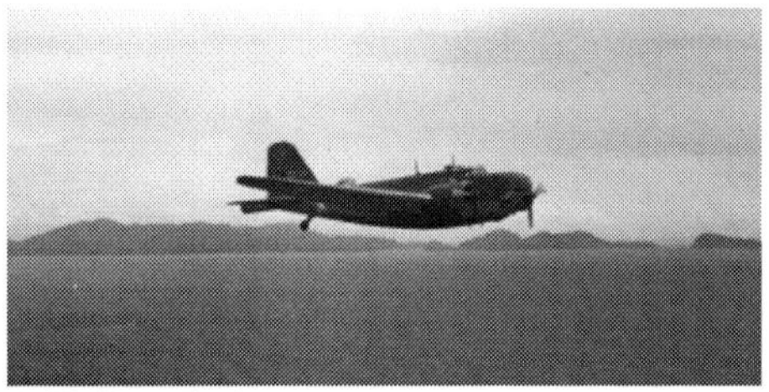
B-18 (Douglas)

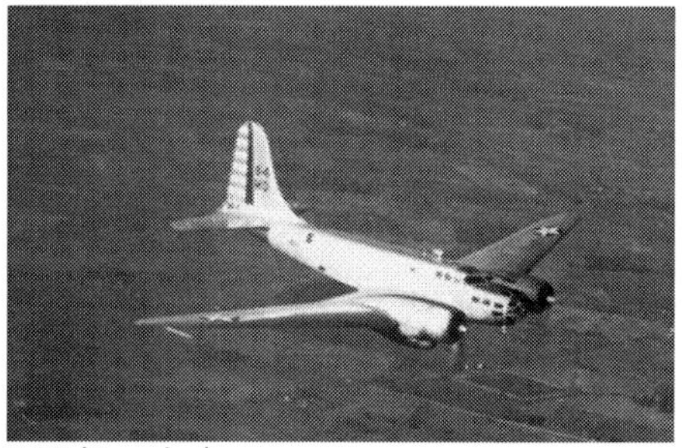
B-23 (Douglas)

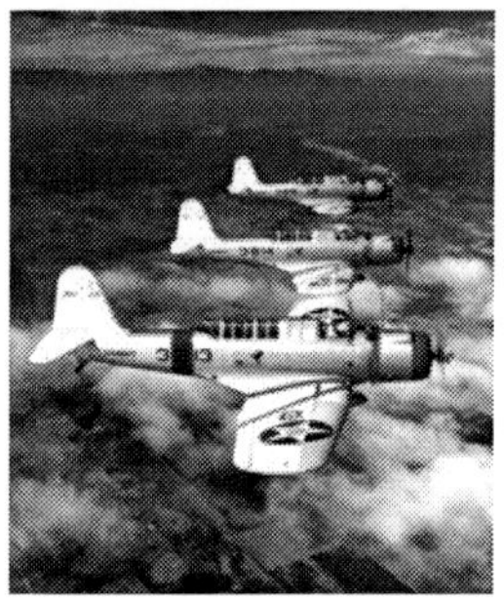
Vindicator (U.S. Navy)

Bombers – Some bomber designs did not go into production or were not used extensively by American airmen. The Buccaneer and Vindicator were flown by the British. The YB-35 never went into production while the B-18 and B-23 were not produced in significant numbers because of design shortfalls.

About the Author

John is a skilled aviation historian focused on mid twentieth-century general aviation, military aircraft and the people that made them possible. As a member of AOPA (Aircraft Owners and Pilots Association), EAA (Experimental Aircraft Association) and an honorary member of the Danbury, Connecticut WWII Lost Squadron Veterans organization, his stories are recognized for their excellence in accuracy and authenticity.

He has written over 200 aviation articles covering everything from Autogiros, Blimps, Corsairs, Mustangs and secret B-24 missions to bomb Tokyo, to Red Bull Air Race pilots and the stories behind private aircraft. His stories have been published in more than 20 publications. John also writes on social media and web marketing topics. John is an American Marketing Association Silver Effie winner and has spoken to large and small audiences in over 60 countries around the world.

This is John's fourth book. It and other books he publishes are supported by the large collection of early aviation press photographs he has collected through the years. questions@vintageflyer.com

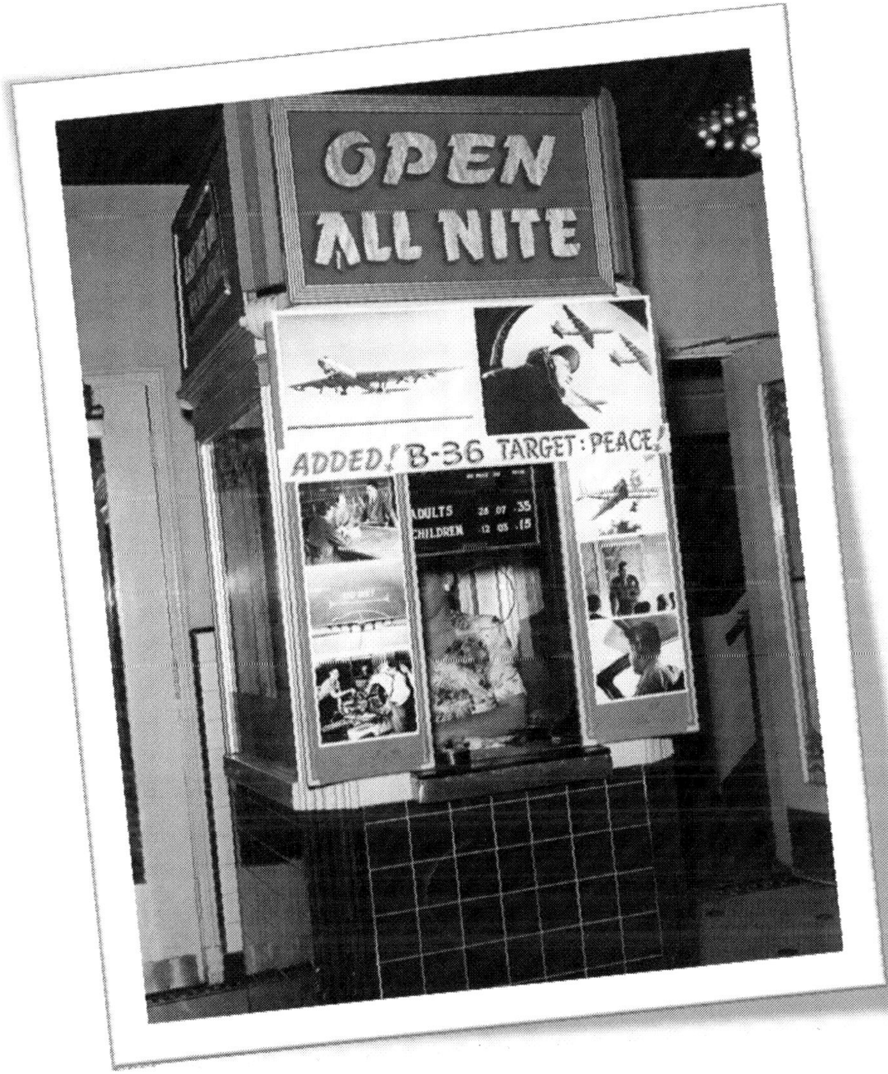

Target Peace – Target Peace was a ten minute informative film about the B-36 Bomber and how the aircraft was helping to keep peace in the world. Several movie theaters carried the film as an extra to run with feature films. (U.S. Air Force)